Oil and Gas Drilling Funds

A Primer for Attorneys and Investors

OIL AND GAS DRILLING FUNDS

A PRIMER FOR ATTORNEYS AND INVESTORS

Grayson Cecil Fox

State Bar of Texas
Created in 1939
Professional Development Program

Austin 1983

The State Bar of Texas, as part of its Professional Development Program, publishes practice books prepared and edited by knowledgeable authors to give practicing lawyers and judges as much assistance as possible. The competency of the authors ensures outstanding professional products, but of course neither the State Bar of Texas, the contributors, nor the editors make either express or implied warranties with regard to the use or freedom from error of this publication. In the use or modification of these materials, each lawyer must depend on his or her own knowledge of the law and expertise.

The use of the masculine gender throughout this publication is purely for literary convenience and of course should be understood to include the feminine gender as well.

International Standard Book Number: 0-938160-32-X
Library of Congress Catalog Card Number: 83-60327

Austin, Texas 78711

Printed in the United States of America

STATE BAR OF TEXAS

COMMITTEE ON CONTINUING LEGAL EDUCATION

1982–1983

About the Author

The author, Grayson Cecil Fox, returned to law school at the University of Texas after ten years in the field of mental health administration in Houston. She received her undergraduate degree from the University of Texas in Austin in Economics and later obtained a Masters in Public Health from the University of Texas School of Public Health in Houston.

She is the daughter of Mary Reed Cecil and the late Judge Lamar Cecil of Beaumont and is married to Fred K. Fox, an inventor and petroleum engineer. Her law practice in Austin is chiefly in the areas of corporate and oil and gas law, and she receives invaluable advice and counsel not only from her husband but also from her three sons, one of whom is in the oil business in Houston and two of whom will graduate from the University of Texas with degrees in Petroleum Land Management and Geology.

Book Publishing Division
Professional Development Program
State Bar of Texas

Susannah R. Mills, *Editor in Chief*
Franklin Holcomb, *Legal Editor*
Nora Kronenthal, *Legal Editor*
Irene Aldrich, *Chief Production Editor*
Deborah S. Speas, *Production Editor*

Contents

List of Illustrations

STATE BAR OF TEXAS

ORRIN W. JOHNSON

OFFICE OF THE PRESIDENT
402 E. Van Buren Street
Harlingen, Texas 78550

The State Bar of Texas proudly presents Oil and Gas Drilling Funds: A Primer for Attorneys and Investors, which will prove an indispensable guide to practitioners advising clients about the intricacies of setting up and marketing drilling funds.

In this concise handbook, author Grayson Cecil Fox outlines basic facts about the workings of the oil and gas industry, examines factors to be considered in choosing the form of business entity to be used for the drilling fund, and clarifies many of the complex tax and securities law aspects of creating these funds and attracting investors.

Providing such helpful practice aids for Texas lawyers is an important mission of the State Bar of Texas, and we are deeply grateful to the authors and other contributors who so generously give their time and knowledge for the furtherance of this mission.

Orrin W. Johnson
President, State Bar of Texas

Acknowledgments

When research for this book began, I was struggling to comprehend my first oil and gas drilling fund prospectus in order to advise my mother about a potential investment. Having rendered to her my best professional advice of "who knows?" I resolved to look into the matter. This book is the result and is intended not only for investors but also for attorneys confronting their first drilling fund. Inclusion of the word *primer* in the title is meant to dispel any claim that this is the definitive work. My goal will have been met if readers become more comfortable in considering oil and gas drilling funds for investment and if attorneys can use this book to provide broad guidelines for preparing their first fund.

Two people deserve special recognition for the many hours they spent giving me a broad background in the oil and gas business; they are Fred Kerwin Fox and Robert A. Shepherd, Jr. While I reserve to myself responsibility for any discrepancies and inaccuracies in this book, I acknowledge that any scholarship herein found results from the invaluable assistance and review provided by University of Texas law professors Barbara B. Aldave, Parker Fielder, Robert Hamilton, and Ed Turley and by Karen J. Bartoletti of Graves, Dougherty, Hearon & Moody, Austin.

G. C. F.

Oil and Gas Drilling Funds

A Primer for Attorneys and Investors

CHAPTER ONE

Introduction

§ 1.01 General Considerations

There is a certain esoteric nature to securities transactions in general. In normal merchandise transactions, the purchaser examines the item to be purchased and, based on firsthand knowledge, decides if the quality and the price of the item are satisfactory. But a certificate representing a security gives no information unless something is known about the underlying business that it represents. To analyze the underlying business, a potential investor of a security may turn to the prospectus. This effort may succeed, but often an interested investor will be defeated by the form of the prospectus itself: "Masters of this medium [prospectus writing] utilize turgid prose to enshroud the occasional critical revelation in a morass of dull, and—to all but the sophisticates—useless financial and historical data."[1]

The difficulties surrounding the investor who pursues a critical analysis of any business are multiplied when he is faced with the complexity and specialized nature of the oil and gas business, particularly when unscrupulous promoters seek to take advantage of the situation. (In the common parlance of the oil business, an investor is referred to as a "mullet," which is a small fish that swims blindly in the Gulf of Mexico.)

Investment in an oil and gas drilling fund, despite its complexity and the possibility of abuse, has the potential for tremendous return on investment coupled with substantial tax incentives; but the layman may hesitate to invest in such a venture because of his lack of knowledge about the oil and gas business, compared, for example, to his general understanding of the real estate business. Baroid Corporation recently compiled the statements of elementary schoolchildren who were asked about the oil business, and the following statement reflects the confusion that can begin at an early age: "One thing I don't know about drilling for oil is plenty."[2] Even the experts cannot say with any certainty what is beneath the ground, and so the risks of an investment in an oil and gas drilling fund are substantial.

§ 1.02 Purpose and Scope of Book

Analysis of a drilling fund prospectus or of an offering memorandum is a complicated task. Furthermore, if one invests in a drilling fund, there are progress reports to analyze or calls for additional assessments to consider.

---

1. Feit v. Leasco Data Processing Equip. Corp., 332 F. Supp. 544, 565 (E.D.N.Y. 1971).

2. *Wildcatters Look Something Like People*, 19 BAROID NEWS BULL. 4.

Without an understanding of the underlying business, the stakes are too high and an investor very sensibly hesitates to invest. This book will attempt to sketch, in broad outline, the oil and gas business and its investment opportunities to provide the background necessary for an investor or a general practitioner of law to understand drilling funds. Chapter 2 contains basic discussion of geology, ownership of oil and gas, and drilling and producing oil and gas wells, and the structure of drilling funds is analyzed in chapter 3. Later chapters deal with the tax and securities aspects of these funds. The goal of this book will be met if it enables a reader to competently consider an investment in a drilling fund or to confidently advise another about such an investment.

CHAPTER TWO

The Oil and Gas Business

[*Unassigned section numbers are reserved for possible use in supplementation.*]

I. Introduction

§ 2.01 Purpose and Scope of Chapter

Part of the mystique of the oil business lies in the terminology surrounding it and the fact that few investors have any background that prepares them to feel comfortable discussing the business, much less investing in it. Misconceptions are rampant, as another quote from one of the schoolchildren indicates: "Petroleum can only be found under ground at the present time!"[1] The next four parts of this chapter will describe the basic features of the oil and gas business, including geology, ownership of oil and gas, and drilling and producing of wells.[2]

II. Geology

§ 2.11 Creation and Types of Petroleum Reservoirs

Petroleum is a generic name for combustible hydrocarbons, and the molecular structure varies from the simplest, methane (CH_4), to much more complicated chains of hydrogen and carbon atoms, such as octane (C_8H_{18}). It can occur in gaseous, liquid, or solid state. The generally accepted theory is that the formation of hydrocarbons derives from the decay of organic matter—plant and animal life—and from the subsequent pressure and temperature applied by the formation of sedimentary rock during geologic eras; thus, the surface of the earth is not fixed. Organic matter decays, rocks are shattered by erosion, and the residue is washed into the sea. This sediment builds up over the ages, and the layers are subjected to increasing pressure. Hydrocarbons are formed from the organic matter thus compressed, and sedimentary rock is formed from the compaction of fragmentary materials. The same pressure that creates these substances then forces the hydrocarbons into the

1. 19 Baroid News Bull. 4.

2. *See generally* M. Ball, This Fascinating Oil Business chs. 1–9 (1963); S. McDonald, Petroleum Conservation in the United States ch. 2 (1971); Oil and Gas Production, Interstate Oil Compact Commission (1951); 1 H. Williams & C. Meyers, Oil and Gas Law chs. 1–3 (1959).

porous sedimentary rock. Later, changes in structure in both the sea and the land cause shifts in the crust of the earth, resulting in uplifts and degradations, and what was once seabed becomes land again.

Some rocks are not formed from sediment but have solidified from a molten state. These are called igneous rocks, of which granite is a common example. Extreme variations in heat and pressure can change the composition of igneous and sedimentary rocks and rearrange their mineral constituents, resulting in metamorphic rocks, so named because these rocks have metamorphosed; thus, shale becomes slate and limestone becomes marble.

Sedimentary rocks are of great importance to the petroleum geologist because they are the major reservoir rocks for hydrocarbons. Common forms are sandstone, limestone, and shale. The igneous and metamorphic rocks act as the "basement" structure on which the oil-bearing "sands" are deposited. Thus, the crust of the earth comprises layers of these three types of rock.

Sedimentary rock is found at many different levels. If a layer of sedimentary rock outcrops at the surface, the layer will often take the name of the locale in which it appears. Hence we have such oil-producing formations as Austin Chalk (a limestone) and Frio Sand (a sandstone) in Texas and Wilcox Sand in Oklahoma. The same geologic layer may have a different name in another area.

The sedimentary rocks are very porous and have a high water content. Oil and gas tend to migrate with the water through the rocks until the movement is checked by an impervious structure. When such an obstacle is encountered and a pool or reservoir is formed, the gas will tend to form the top layer, followed by the oil, and the water will constitute the lowest layer.

Hydrocarbons are found in two general kinds of traps: structural and stratigraphic. Examples of structural traps are anticlines (figure 2.1) and faults (figure 2.2). An anticline results when a gentle uplift produces an arch in the stratum that was probably level originally. Faults are caused by more violent upheavals that shift and break the rock strata. On the Gulf Coast the slope of the landmass southward eventually caused it to break and crack along fault lines.

In a stratigraphic trap, the monoclinal structure "tightens up" and loses its porosity and permeability, or "pinches out," between two planes of impervious rock. The East Texas field, which is the largest oil field in the forty-eight contiguous states with a capacity of approximately five billion barrels of recoverable oil, is an example of a stratigraphic trap (figure 2.3). Another structure common to the Gulf Coast is the salt dome, which produces both structural and stratigraphic traps (figure 2.4).

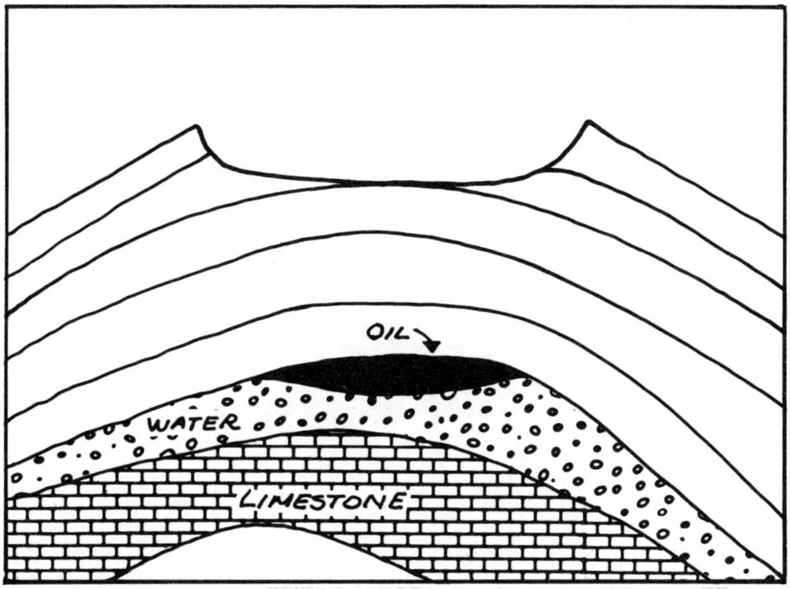

Figure 2.1. Anticline

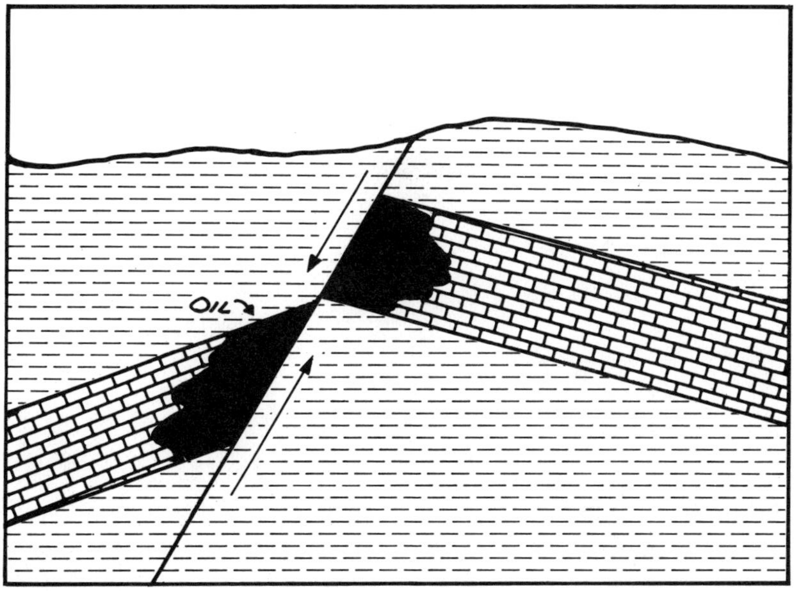

Figure 2.2. Fault

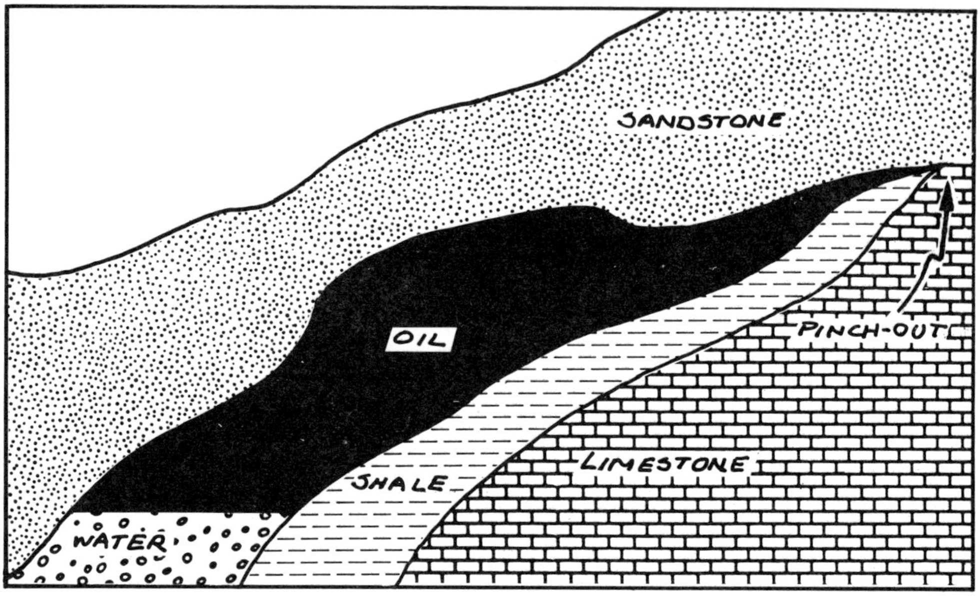

Figure 2.3. Stratigraphic trap; East Texas field

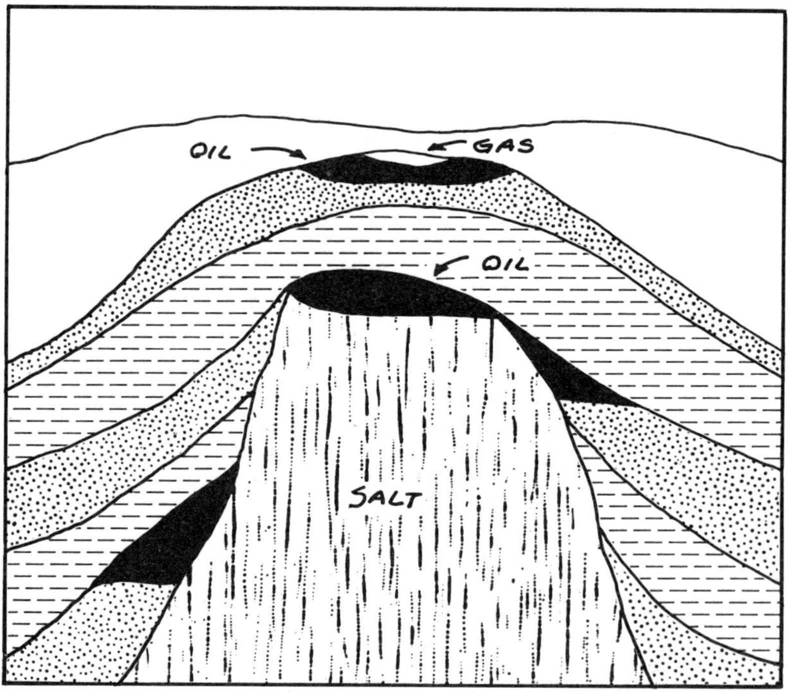

Figure 2.4. Salt dome

§ 2.12 Locating Petroleum Reservoirs

The petroleum geologist's job is to locate structural and stratigraphic traps. Usually the technology available is more successful in locating faults, anticlines, and salt domes than it is in finding stratigraphic traps, which are more often discovered by actual drilling in an area. A geologist will use a number of methods to construct a contour map of the subsurface of the earth. A first task is to examine surface formations. The Fort Collins field was discovered in Colorado by observations that the normal slope of the land eastward, away from the mountains, changed in a lake bank and dipped westward, indicating an anticline. Also, because the crust of the earth forms an imperfect seal and hydrocarbons below may seep to the surface, soil analysis can indicate the presence of a reservoir of oil and gas.

Magnetic forces can also be used to discover the broad contours of the subsurface. Because each of the three classes of rock has a different iron content, each class also differs in magnetic attraction. Igneous and metamorphic rock often contain more iron, and thus a magnetometer (a device to measure magnetic intensity) can be used to plot anticlines and domes, because the rock with the greatest pull will be nearest the surface.

Seismographs were originally developed to measure earthquakes but are now also used to discover petroleum reservoirs. An explosion sends out shock waves, and the seismograph measures the sound waves reflecting back through the rock layers. The waves reflected back indicate the velocity of sound in each subsurface layer. These varying reflections and velocities are combined to delineate subsurface structures.

Employing any of these means or combinations, the geologist has a contour map drawn, with each line representing every point that is at the same elevation (figure 2.5). A standard joke in the oil business is that a high-rolling

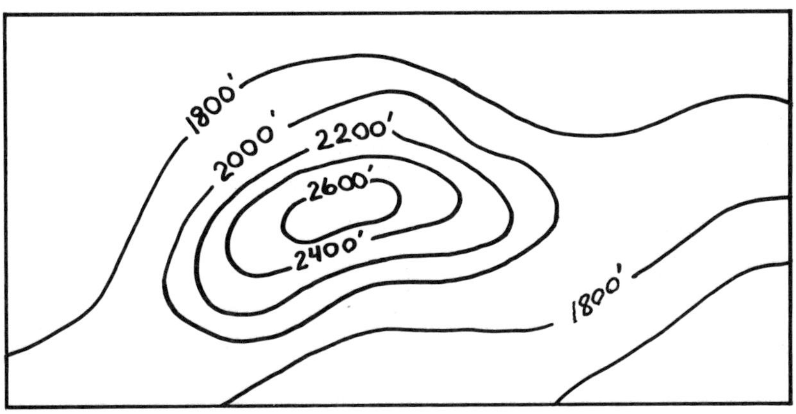

Figure 2.5. Contour map

sponsor will sometimes produce a contour map by tracing the contours of his oak office door!

Armed with all the available information, the geologist is now ready to recommend where a well should be drilled. The next problem encountered is acquiring the right to drill in the location selected.

III. Ownership of Oil and Gas

§ 2.21 General Considerations

At the simplest level, the owner of land in fee simple absolute is said at common law to own the land to an indefinite extent upward as well as downward. Thus the landowner possesses the exclusive right to explore for oil and gas, subject only to the police power of the state and the correlative rights of neighbors. The landowner may grant any part of the right to explore for oil and gas to another or may convey the surface to another and reserve any part of the exploration rights. The interest granted or reserved may be a mineral interest, a royalty interest, or a leasehold estate. Texas law holds that the landowner's interest in the oil and gas beneath the land is one of absolute ownership in place.[1] This ownership is subject to the law of capture, which means that, although the landowner may extract oil and gas from a reservoir that underlies his own land as well as that of a neighbor, the neighbor has a correlative right to do the same thing. The one who "captures" the oil first owns it.[2]

§ 2.22 Types of Interests

(a) Mineral Interests

A mineral interest is created by a mineral deed or a reservation. The interest may be an entire mineral interest or an undivided interest; it may be in fee simple, in fee simple defeasible, or for life. The owner of an entire mineral interest, if different from the owner of the surface interest, has the right with

1. Stephens County v. Mid-Kansas Oil & Gas Co., 113 Tex. 160, 254 S.W. 290 (1923).

2. *Id.* 254 S.W. at 292; *see also* Hardwicke, *The Rule of Capture and Its Implications as Applied to Oil and Gas*, 13 Tex. L. Rev. 391, 405 (1935).

certain limitations to go on the land and explore for oil and gas.[3] If the interest is an undivided one, it is characterized as coownership.

(b) Royalty Interests

A royalty interest may also be created by deed or by reservation but carries with it no right to go on the land and drill for oil and gas.[4] The owner of the royalty interest is, however, entitled to share in the proceeds from drilling operations.[5] Generally, a royalty interest is an incident of an oil and gas lease; that is, the owner of the mineral estate leases the minerals for drilling purposes and reserves a royalty interest that usually ranges from one-eighth to one-fourth. An example of a royalty interest is: *A*, the owner of the mineral estate, leases the minerals to *B* and retains a ⅛ royalty interest. *A* may then sell ½ of his royalty interest to *C*, and then *A* and *C* each has a 1⁄16 (½ of ⅛) royalty interest. This interest does not bear the costs of drilling but does normally bear some costs, such as production and severance taxes, transportation charges, and expenses to make the product marketable (compression of gas or extraction of other substances).

(c) Leasehold Interests

A leasehold interest conveys the authority to go on the land and explore for oil and gas[6] and typically is stated in terms of years "and so long thereafter" as minerals are produced in paying quantities. A bonus is paid to the owner of the mineral estate upon execution of the lease, and the term of the lease is usually for five years. However, if drilling is not begun within the first year, a delay rental is usually required at each interval of one year during the primary term of the lease. The lease form most often used is that called a "Producer's 88." The owner of the mineral interest should know that this term does not imply a standard lease form. The term "Producer's 88" is of general use in the oil and gas business, and leases so identified can take many different forms, some of which are considerably less advantageous to landowners than others. Generally, a lease provides for certain expressed cove-

---

3. Marrs v. Railroad Comm'n, 142 Tex. 293, 177 S.W.2d 941, 948 (1944); *see* 1 H. Williams & C. Meyers, Oil and Gas Law, §§ 217, 218 (1980), for discussion of conduct by oil and gas operator that is injurious to the surface estate. *See generally* Annot., 146 A.L.R. 880, 899 (1943), for effect of severance.

4. 1 H. Williams & C. Meyers, *supra* note 3, at § 303.4.

5. Carroll v. Bowen, 180 Okla. 215, 68 P.2d 773 (1937).

6. Barwise v. Sheppard, 299 U.S. 33, 57 S. Ct. 70, 81 L. Ed. 23 (1936).

nants, such as the covenant to offset if a well is brought in on adjacent property, while other covenants can be implied. For instance, a lessee is held to impliedly covenant to reasonably develop an oil and gas lease[7] and to market the oil and gas in a reasonable time.[8] Further covenants to operate with care have also been implied, and lessees have been found to abridge this covenant by careless conduct,[9] by premature abandonment of drilling efforts,[10] and by failure to use advanced production techniques.[11]

(d) Working Interests and Farmouts

The owner of a leasehold interest is said to hold the working interest, which bears the costs of drilling and producing the wells. The owner of the working interest may in turn execute a "farmout" agreement, giving another operator the right to drill. The owner of the working interest may then retain an overriding royalty interest. An example of the creation of an overriding royalty interest is: *A* leases to *B* and retains a ⅛ royalty. *B*, owner of 100 percent of the working interest and ⅞ of the production, farms out the lease to *C* to drill and retains a ⅛ overriding royalty interest (⅛ of ⅞). *C* then owns 100 percent of the working interest and 49⁄64 of the production (⅞ of ⅞). Often a producing well is owned by a combination of these interests, and, when the oil and gas are sold, the purchaser will require that a division order be executed to establish the various ownership percentages.

(e) Top Leases

One final lease form of interest is the top lease. This lease is superimposed on an existing lease and is automatically effective if the underlying lease should lapse because of failure to drill in the required time, because of failure to pay delay rentals, or because production ceases after the primary term.

§ 2.23 Acquisition of Drilling Rights

The leasing agent, commonly known as the landman, has the job of discovering the ownership of the oil and gas interests in the area where the geologist

---

7. Waggoner Estate v. Sigler Oil Co., 118 Tex. 509, 19 S.W.2d 27 (1929).

8. Howerton v. Kansas Natural Gas Co., 81 Kan. 553, 101 P. 47, *on rehearing*, 82 Kan. 367, 108 P. 813 (1910).

9. Guth v. Texas Co., 155 F.2d 563 (7th Cir. 1946).

10. Empire Oil & Ref. Co. v. Hoyt, 112 F.2d 356 (6th Cir. 1940).

11. Wadkins v. Wilson Oil Corp., 199 La. 656, 6 So. 2d. 720, 721 (1942).

has determined that oil and gas reservoirs exist and of acquiring the right to drill either by lease or by acquiring a farmout. This result accomplished, the operator is ready to drill.

IV. Drilling the Well

§ 2.31 Types of Wells

An area that a geologist chooses for drilling in an unproven territory is called wildcat territory, and the wells drilled will be termed exploratory or wildcat wells. If the area has already been proved, the well is called developmental and can be further described as either a close-in well or a step-out well (figure 2.6).

§ 2.32 The Drilling Process

First, the roads to the drill site are constructed and the drilling rig and derrick are assembled. The drilling then begins, usually by means of a rotary

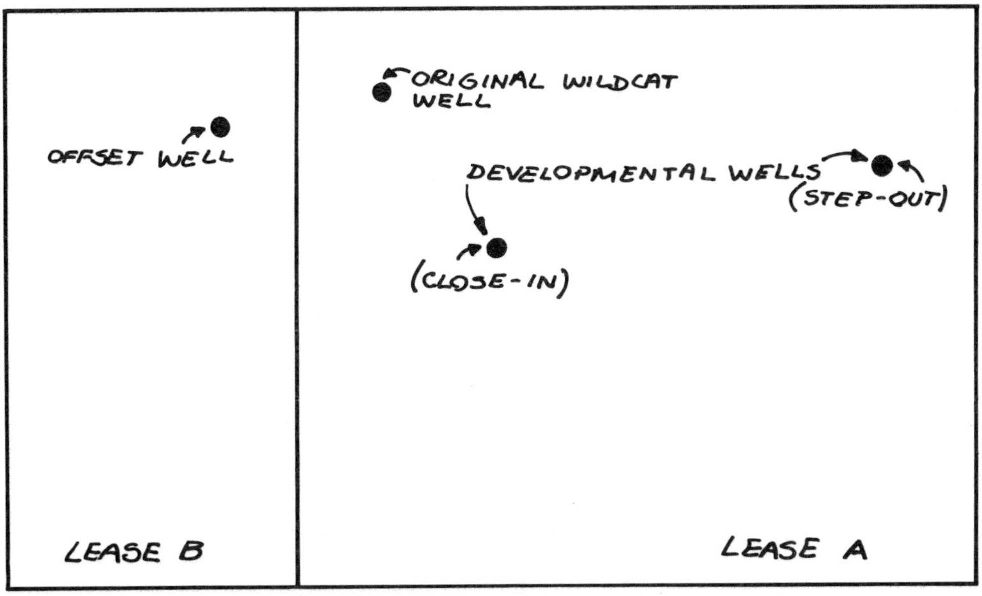

Figure 2.6. Exploratory (wildcat) and developmental wells

drill.[1] The drill pipe, which comes in lengths of approximately thirty feet, is rotated from the derrick floor, and a bit is attached to the bottom of the drill stem, cutting through the rock formation as the pipe rotates. As drilling continues, successive joints of drill pipe are attached from the derrick floor by securing the pipe with heavy tongs and screwing one section to another. The lower sections of the drill stem are called the drill collar and are often grooved in a spiral configuration to prevent sticking. When the bit wears out, the entire drill stem must be pulled from the hole, taken apart in two- or three-joint lengths, and a new bit attached. Raising the entire drill stem from the hole is called "making a trip."

Drilling mud is pumped down through the drill stem and bit during drilling, returning to the surface via the outside of the drill stem bearing with it the cuttings from the hole. The drilling mud is also useful in many other respects: it cools the bit, it controls the pressure within the hole, and it seals off any porous strata that have been drilled. At various intervals the drill stem is removed and an electrical logging device (generally called a Schlumberger for the man who invented it) is lowered to measure the layers of rock and record the porosity and, it is hoped, the oil- and gas-bearing content.

After several thousand feet have been drilled, casing is inserted into the hole. This surface casing protects freshwater zones from contamination during drilling and prevents the sides of the hole from caving in. Casing comes in joints of thirty to forty feet, and each joint is separately attached as the casing is lowered into the hole. Successive strings of casing are often used, and each string normally extends from the surface of the well down (figure 2.7). If oil and gas are discovered, production casing is cemented into place (figure 2.8). Thus the well proceeds with the sequence of drill, log, and case until the target sands are reached.

§ 2.33 The Drilling Crew

The drilling crew consists of a number of different people. The tool pusher ordinarily lives in a trailer on site for the duration of drilling and is in charge of the drilling operation of the rig. The driller manages the operation on the derrick floor and usually works an eight- or twelve-hour shift depending on the location of the well and the availability of crew members. Working under

1. Cable-tool drilling is used in approximately 20% of the wells drilled nationally, though few are drilled with this technique on the Gulf Coast. With this method the hole is punched into the earth by raising and lowering a heavy bit suspended from a cable.

the driller are roughnecks, who work on the derrick floor and maneuver the heavy tongs that are used to attach the various pipes. Roustabouts perform miscellaneous chores off the derrick floor.

§ 2.34 Preparation for Production

If the geologist is correct and oil and gas are found, the production string of casing is set to extend through the oil- and gas-bearing formation. The casing is then perforated and the flow of oil begins. If the well does not flow easily at first, various methods are employed to increase the flow, such as "acidizing" the well with hydrochloric acid or "fracturing" the well with hydraulic pressure to increase the flow capacity of the formation. The derrick is then dismantled and a "Christmas tree" is installed, which is a control head that fits on the innermost string of casing and has valves regulating the flow of oil and gas.

§ 2.35 Drilling Costs

The cost of drilling a well increases exponentially with the depth of the well. The deepest well drilled thus far is approximately 35,000 feet, while the average well is 6,000 feet. An onshore well of 2,000 feet can cost approximately $40,000, and a well of 16,000 feet can cost $3.5 or $4 million. Offshore drilling is substantially more expensive. Most of the costs of drilling a well are referred to as intangible drilling costs. See chapter 4 of this book on the tax treatment of these costs.

§ 2.36 Drilling Technology

As indicated in figures 2.7 and 2.8, the vertical scale is greatly reduced. All drawings of oil wells tend to obscure the fact that thousands and thousands of feet of pipe are being rotated from the derrick floor. If a picture of a derrick is one-half inch high, the corresponding picture of the drill stem should extend eight inches to depict a shallow well, as illustrated on the opening chapter pages of this book. Because of the length of the drill stem, drilling is limited by the tensile strength of the pipe, which is severely strained by the torque created in the rotary process. When the drill pipe is at the bottom of a 15,000-foot well, the weight of the entire string rests on the top joint and can be more than 100 tons.

There is now a major effort to develop a "down-hole" drill that operates

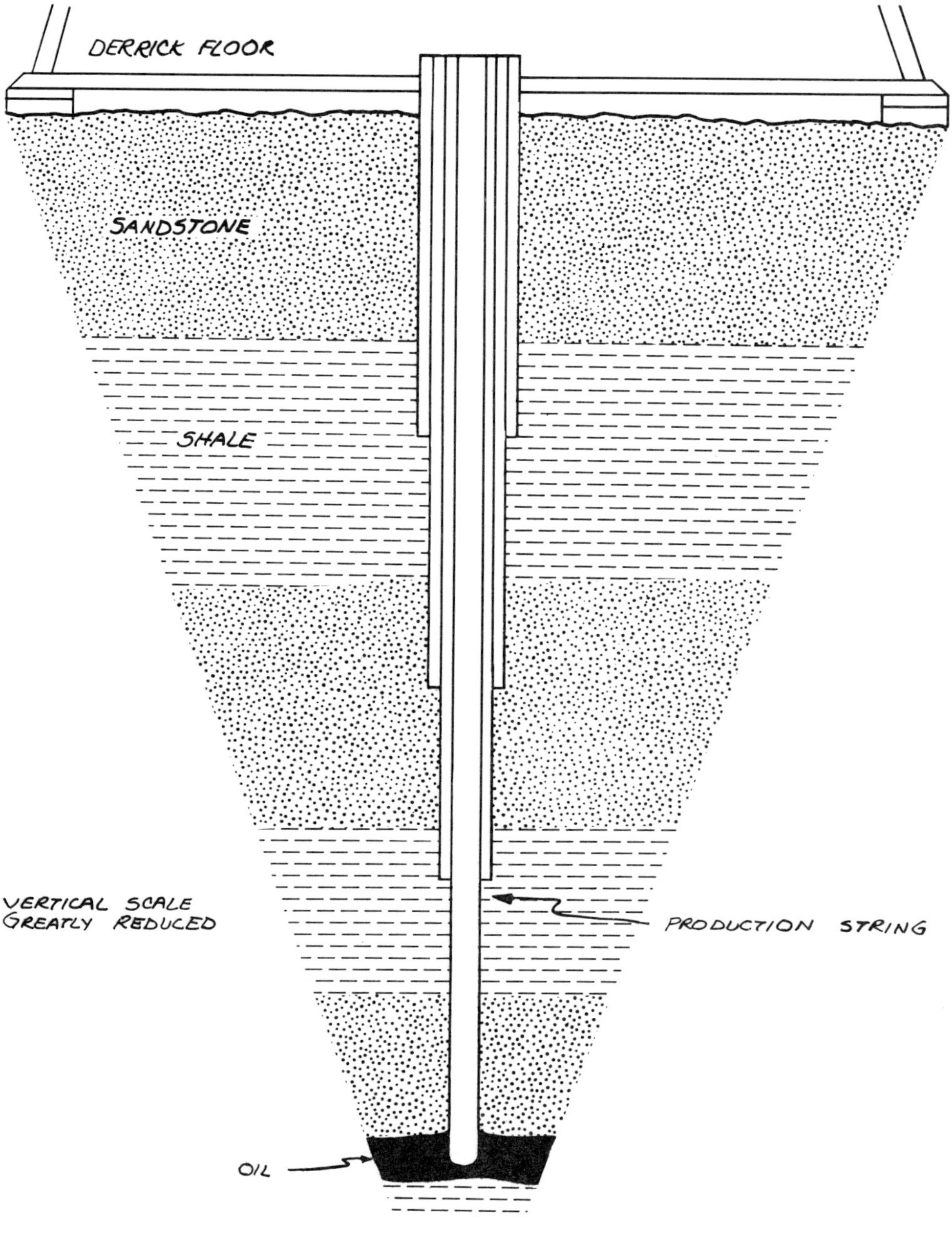

Figure 2.7. Casing

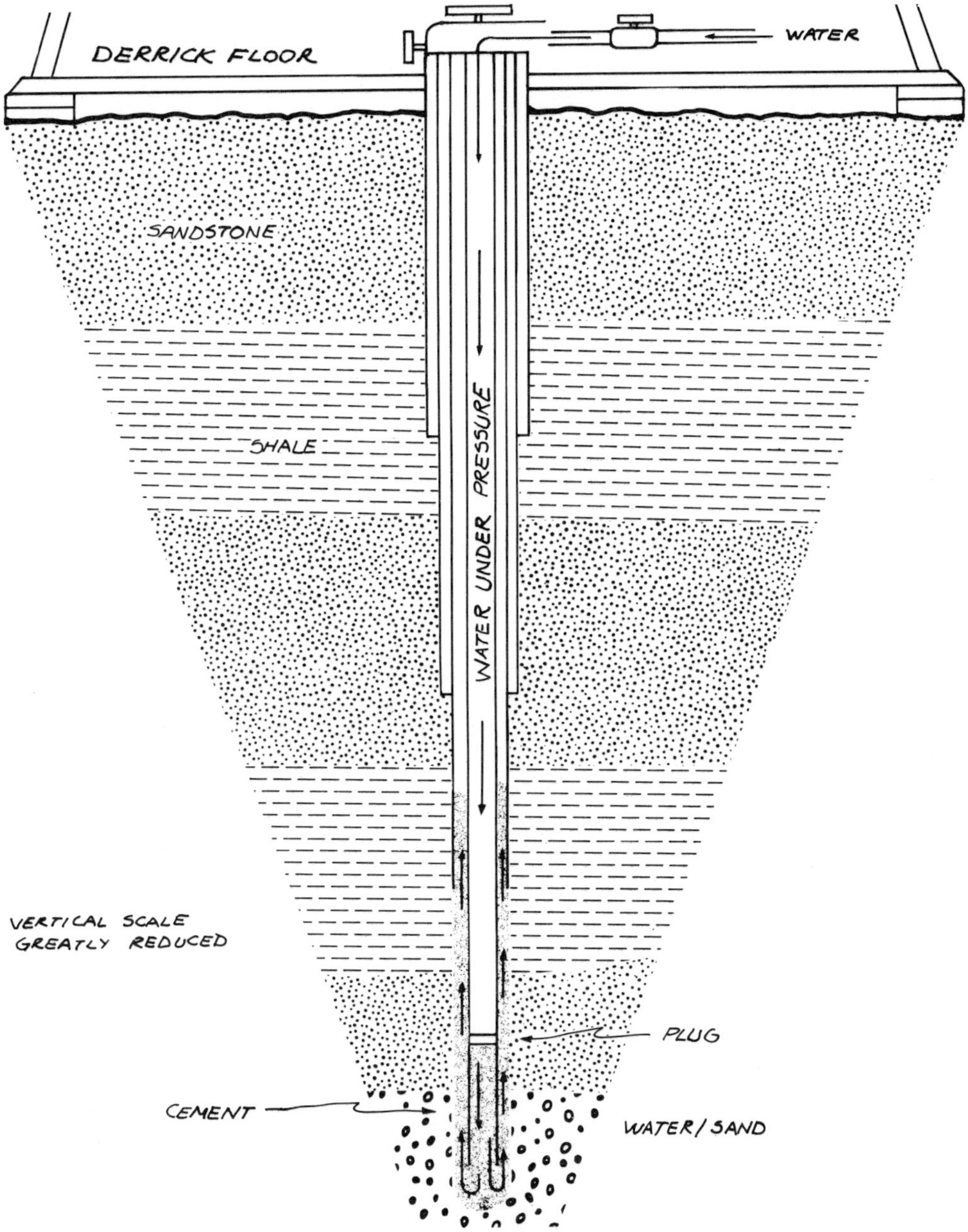

Figure 2.8. Cementing

at the bottom of a stationary drill stem; its turbine motor is powered by the force of the mud flowing through it.[2] Such a motor allows drilling speeds of 800 rpm as compared to a rotary drilling speed of 150 rpm. The down-hole motor is particularly advantageous for drilling deep wells because the problems of torque created by a rotary drill are eliminated and because the increased drilling speed greatly reduces the drilling time and hence the cost of a well.

V. Producing the Well

§ 2.41 Reservoir Engineering

Because oil and gas in a reservoir are under great pressure, they will expand if a well is completed in the reservoir; the pressure on the oil and gas is reduced by the well, and this reduced pressure will cause them to flow toward the hole. If a well is completed in the gaseous area of a reservoir, not only will the free gas escape, but also the gas that is in solution with the oil will escape, bringing some of the oil to the surface but leaving the remaining oil heavier and less mobile with the result that it will not flow easily into the well bore. On the other hand, if the well is completed in the oil-bearing part of a reservoir, the pressures are lowered in the oil section of the reservoir, which promotes a more efficient movement of oil toward the well bore. The pressure of any water lying below the oil in the reservoir may also act to push the oil and gas toward production. Older reservoir-engineering theory considered ultimate recovery to be sensitive to the rate of production and held that for each reservoir there was some maximum efficient rate (MER) of production at each step of depletion. The MER was thus considered functionally dependent on the number of wells in the reservoir, and often spacing of wells was mandated in an effort to maintain production in a field. Newer theories of reservoir engineering conflict with the traditional MER concept and postulate that the rate of production does not affect ultimate recovery.

---

2. Moore, *Drilling Technology: A Hard Look at Downhole Motors*, Oil & Gas J. 75 (Mar. 9, 1981).

§ 2.42 Recovery

The natural drive of the gas and water in a reservoir can sometimes provide recovery of 75 percent of the oil in place if utilized properly. When natural drive is almost exhausted, secondary recovery can be obtained by injection of water and gas into the oil zone through adjacent wells. Another method of recovery (tertiary) is the injection of steam to heat the oil zone and thereby reduce the viscosity of the oil. Operating on a field-wide basis rather than on a lease-by-lease basis can greatly increase secondary and tertiary recovery. In the final stages a well often produces no more than ten barrels a day; at that point such a well is termed a "stripper well."

§ 2.43 Storage and Transmission

Oil, when produced, can be stored on site in tanks and transported to market in trucks. Gas transmission, on the other hand, depends on the availability of a pipeline, which is often too expensive to build to a marginal discovery, and so such a marginal gas well is plugged. Much technology is needed to develop methods of storing gas on site as well as transporting it without pipelines. Also, for maximum efficient production of gas, technology is needed to better evaluate potential gas production from a reservoir so that the cost benefit of extending a pipeline can be more accurately determined.

§ 2.44 Production and Price Controls

Various federal and state agencies control the price and production of oil and gas. In Texas the Texas Railroad Commission issues permits to drill and may establish allowable rates of production, control spacing of wells, and otherwise act to conserve natural gas and oil.[1]

Federal regulatory agencies are the Department of Energy, the Economic Regulatory Administration, and the Federal Energy Regulatory Commission. Before 1979, Department of Energy regulations defined three price categories of crude oil. Beginning in April 1979 these regulations began to be phased out, and the last controls expired January 27, 1981. Whereas the posted price for Texas, Louisiana, and New Mexico was $38 a barrel in April 1981, the price in March 1983 was approximately $30 a barrel. The recent

---

1. TEX. NAT. RES. CODE ANN. tit. 3, Oil and Gas (Vernon 1978).

combination of a decrease in the demand for oil and an inability of the Organization of Petroleum Exporting Countries (OPEC) to control production has caused many observers to forecast an even lower price for oil.[2]

Most gas prices continue to be regulated,[3] and in March 1983 "new gas" was selling for $3.34/MMBtu[4] while the price of deregulated gas (that discovered below 15,000 feet) continues to fluctuate, having reached a high of $10.76/MMBtu in 1981 and falling to the $4.00–$5.00/MMBtu range in early 1983. Partial decontrol for "new gas" is scheduled for 1985.[5] Presently, the decrease in the demand for natural gas in the United States has caused Congress to reconsider the entire issue of gas deregulation. Some members of Congress advocate total deregulation now, arguing that market forces will keep the price down in a period of oversupply. Others advocate increased controls and the abolition of "take or pay" gas contracts that require gas purchasers to take a minimum amount of gas each month or pay for it that month and recoup the gas later.

---

2. Robinson, *Much of the Red Ink in Governor's Proposal Is Due to Oil Industry Problems*, Houston Chronicle, Mar. 13, 1983, at 14, sec. 1. The Texas Comptroller of Public Accounts is responsible for estimating anticipated revenue available for each biennial budget, and, because 27.4% of state tax revenue in Texas comes from taxes on oil and gas, forecasting of energy prices is a critical function of that office. The Comptroller has recently predicted an oil price of $28.24 per barrel in 1984 and $29.61 in 1985.

3. Natural Gas Policy Act, 15 U.S.C.A. § 3301 *et seq.* (West Supp. 1982).

4. One MMBtu is equal to approximately 1,000 cubic feet (mcf).

5. 15 U.S.C.A. § 3331 (West Supp. 1982).

CHAPTER THREE

Drilling Funds

[*Unassigned section numbers are reserved for possible use in supplementation.*]

I. Introduction

§ 3.01 General Considerations

Because the capital costs of drilling for oil and gas are enormous, neither independent oil operators nor small oil companies can afford to drill without an infusion of outside capital. A tax-shelter oil investment raises the necessary capital and at the same time frees the sponsor's limited capital. The sponsor is happy to trade certain tax advantages for the opportunity of spreading the risks of the venture while maintaining control of the operation. Increasing the pool of capital also allows the sponsor to conduct a more ambitious exploration program in a given area. Because most of the drilling costs are borne by the investor, the sponsor's risk of losing a large amount of capital in a dry hole is reduced.[1]

From the investor's standpoint, the stakes are high and the risks are great. Despite enormous advances in locating and drilling for oil and gas, an investment in a drilling fund is still very much like a roll of the dice. Investment motivation rests on the potential for returns that far exceeds almost any other type of investment, coupled with tax benefits whether the venture succeeds or not. Before the Economic Recovery Tax Act of 1981 (the 1981 Act),[2] the highest marginal tax rate for individuals was 70 percent. The 1981 Act reduced this rate to 50 percent effective January 1, 1982.[3] The effect of this change on investment in oil and gas drilling funds is not certain, but it seems logical that an investor will analyze the potential profitability of a drilling fund more closely as a result of the 1981 Act, because he will be investing fifty-cent dollars rather than thirty-cent dollars. Thus the tax benefits of the investment are diminished, making the potential for a successful drilling program a more important factor. Investors may also favor real estate investments over drilling funds after 1981 because of the more favorable depreciation schedules adopted by the 1981 Act.[4]

---

1. Humphrey, *Investor Drilling Deals: Observations on Structure, Taxation, and Securities Laws Compliance*, Keplinger Symposium for Oil & Gas Investor 1–3 (1979).
2. Pub. L. No. 97-34, 95 Stat. 172.
3. *Id.* § 101.
4. *Id.* § 141. Rather than using the term *depreciation*, the Act adopts the phrase "accelerated cost recovery system."

§ 3.02 Scope of Chapter

The next part of this chapter will describe the structure and operation of drilling funds. There are many types of oil and gas investments that are possible and equally as many entities into which the investments can be structured. Classification of the entity selected for federal income tax purposes is critical. The charges made by the sponsor, which can vary substantially, as well as some creative ways to analyze risks and obtain leverage, are discussed. Finally, the evaluation of the sponsor as a potential investment partner is considered.

II. Structure of Drilling Funds

§ 3.11 Types of Investments

(a) Public Drilling Funds

The word *public* connotes a fund that is registered with the Securities and Exchange Commission (SEC) in order to raise funds from the public.[1] Chapter 5 of this book deals extensively with the requirements for registration with the SEC. A public drilling fund typically has an oil company as a sponsor, and its size ranges from $1 million to $20 million. Individual units can be purchased for $5,000 to $10,000. A limited partnership form is generally selected, with the oil company acting as the general partner and contributing oil and gas leases that are already owned. The limited partner investors contribute funds to drill and complete the wells. Other drilling funds are called "blind pool" because there are no predetermined leases selected and the general partner may contribute funds just as the limited partner does.

(b) Exempt Drilling Funds

An exempt drilling fund is substantially the same as a public drilling fund except that the former is exempt from registration with the SEC. These exemptions are discussed in chapter 5 of this book, but, generally, funds sold pursuant to an exemption are smaller in size and usually have fewer partici-

1. Bean, *Entity Selection*, 30th Ann. Inst. of Oil & Gas Tax 363, 364 (1979).

pants and larger units of participation. Again, a limited partnership is the usual entity selected.

(c) "POGO" Funds

The "POGO" type of fund is used mainly for financing extensive offshore operations. A fairly large oil company (POGO was the first oil company to do this) will contribute leases to a new corporation that conducts a public offering to raise funds to drill. If successful, the newly formed corporation will merge with a larger oil company, usually the affiliate of the drilling fund corporation.

(d) Promoted (or Carried) Interests

The promoted, or carried, interest represents the classic "oil man's deal." Here an individual oil sponsor has a lease and needs the funds to drill. Three investors are found, each of whom puts up one-third of the costs (which may include leasehold costs) to drill or to drill and complete the first well. In return, the sponsor receives a one-fourth interest. If the first well is good, subsequent developmental wells may be drilled "heads up"—that is, each party, including the sponsor, bears a share of the costs.

(e) Back-in Interests

In the back-in type of investment, the sponsor again owns the lease and needs funds to drill. The lease is assigned to the investors, or funding parties, who pay to drill or to drill and complete the first well. The funding parties are then entitled to 100 percent of the proceeds until they recoup their costs and then usually a substantial (¼ to ½) interest reverts (backs in) to the sponsor, or nonfunding party.

§ 3.12 Investments As Securities

Each of the types of investments described above, including the promoted interest and the back-in interest, qualifies as securities, and therefore the sponsor is required to register with the SEC or seek an exemption.[2] The basic definition of a security as an "investment contract"[3] was interpreted in *SEC v.*

---

2. *See infra* ch. 5.
3. Securities Act of 1933, § 2(1), 15 U.S.C.A. § 77b(1) (West 1981).

W. J. Howey Co.[4] as "a contract, transaction or scheme whereby a person invests his money in a common enterprise and is led to expect profits solely from the efforts of the promoter or a third party."[5] All the types of investments described above appear to fit this definition.

§ 3.13 Choice of Entity

(a) General Considerations

A threshold determination in the creation of a drilling fund is the choice of entity or type of business organization to be used. A limited partnership is the most common choice; because of recent problems in regard to special allocations, however, many sponsors and attorneys are turning to joint ventures that elect to be excluded from the partnership provisions of the Internal Revenue Code.[6] Allocation problems are discussed in chapter 4 of this book. The remainder of this section contains a brief description of the possible organizational forms.

(b) Corporations

Basic characteristics of both a corporation and a partnership are that each has associates and each has a joint profit motive.[7] A corporation is a taxable entity, and presently the top rate for taxation is 46 percent.[8] Ownership of a corporation is generally represented by shares of stock, and the earnings of the corporation are distributed as dividends to shareholders. The dividends are then taxed at the applicable individual rate, thus creating double taxation. The double-taxation problem can be avoided fairly easily by payment of salaries, rent, or interest to the shareholders or by accumulation of earnings with a concomitant increase in the price of the shares, which can be sold after one year and receive long-term capital gains treatment.[9] The more serious problem encountered in using the corporate form for an oil and gas drilling fund is that losses incurred through deductions for intangible drilling costs (IDCs) or percentage depletion cannot be passed through to the shareholders for tax purposes. Because of these problems, a corporation is gener-

---

4. 328 U.S. 293, 66 S. Ct. 1100, 90 L. Ed. 1244 (1946).

5. *Id.* 328 U.S. at 298, 299.

6. 26 U.S.C.A. (I.R.C.) subtit. A, ch. 1, subch. K, § 701 *et seq.* (West 1982).

7. Morrissey v. Commissioner, 296 U.S. 344, 56 S. Ct. 289, 80 L. Ed. 263 (1935).

8. 26 U.S.C.A. (I.R.C.) § 11(b) (West Supp. 1982).

9. *Id.* § 1222, subd. (3) (1982); *see* 19 R. Hamilton, Business Organizations § 33 (Texas Practice 1973).

ally undesirable as a vehicle for a drilling fund since the corporation's most valuable feature, limited liability, can be accomplished in another manner.

(c) Partnerships

The definition of a partnership varies for purposes of state law and federal income taxation. The latter has the broader definition, which encompasses any "syndicate, group, pool, joint venture, or other unincorporated organization."[10] If a partnership is not classified as an association taxable as a corporation (discussed at *infra* section 3.14), it is not a taxable entity; and items of income, expense, loss, and so on, are passed through to the individual partners and taxed only at their respective individual rates.[11] Furthermore, each item retains its essential characteristics in the process; that is, if an income item is capital gain at the partnership level, it remains so at the individual level. Partnerships are easily formed and, within the fairly broad parameters provided by state and federal law, can be structured to suit the needs of the partners.

A general partnership has joint management by all the partners, and any partner may bind the partnership and create joint and several liability for all the partners. A limited partnership has a general partner, who is personally liable, and a number of limited partners, whose liability extends only to the amount of their investment in the partnership.[12] The general partner controls the business of a limited partnership, and a limited partner risks loss of limited liability if he exercises control of the partnership. A limited partnership is thus an ideal form for drilling funds, because the sponsor, who wishes to maintain control of the operations, may do so without providing all the capital.

(d) Joint Ventures

The joint venture form of business organization is generally a joint enterprise, as its name implies, entered into for profit by two or more people to accomplish a particular, time-limited project. Mere coownership of property is not a joint venture or a partnership for tax purposes unless the coowners actively carry on a trade or business.[13] Furthermore, in Texas, state partner-

---

10. 26 U.S.C.A. (I.R.C.) § 7701(a)(2) (West 1967).

11. 5 Fed. Tax Coordinator 2d (RIA) § B1100 (1981).

12. 19 R. Hamilton, *supra* note 9 (§ 3.13(b) herein), at § 211.

13. Treas. Reg. § 1.761-1(a), T.D. 7208, 37 Fed. Reg. 20,686 (1972).

ship law expressly states that "operation of a mineral property under a joint operating agreement does not of itself establish a partnership."[14]

In practice, the distinction between coownership and a joint venture or partnership is difficult to make. If a joint venture is not classified as an association taxable as a corporation, it will be considered a partnership for tax purposes unless it elects to be excluded from the application of subchapter K, which involves partnership taxation.[15] This election may be made if the venture is for the "joint production, extraction, or use of property, but not for the purpose of selling services or property produced or extracted."[16] This exclusion was designed for oil and gas ventures and, as interpreted by the regulations, requires that the members (1) must be able to compute their income separately, (2) must reserve the right to take the oil and gas in kind, and (3) must not jointly sell the oil and gas for a period of more than one year.[17]

Because a joint venture is in essence a narrow-purpose partnership, despite an election out of subchapter K for tax purposes, most state partnership rules will apply irrespective of tax classification.[18] For example, partnership rules of joint and several liability will apply.

In the past, joint ventures were often structured as coownerships coupled with an operation agreement with one owner to manage the project.[19] Because of the possibility of personal liability for all the venturers, this type of drilling fund fell into disuse in favor of limited partnerships.[20] However, there is a considerable revival of interest in joint ventures, which is discussed in chapter 4 of this book in the sections on special allocations.

(e) Subchapter S Corporations

A subchapter S corporation,[21] available for certain small business corporations, has characteristics of both a corporation and a partnership, but the uniquely distinguishing feature is that there is no tax at the corporate level. Recent legislation[22] has removed many of the stumbling blocks encountered

---

14. TEX. REV. CIV. STAT. ANN. art. 6132b, § 7, subd. (5) (Vernon 1970).

15. 26 U.S.C.A. (I.R.C.) § 761(a) (West 1982).

16. *Id.* § 761(a)(2).

17. Treas. Reg. § 1.761-2(a)(1), (3), T.D. 7208, 37 Fed. Reg. 23,161 (1972).

18. 19 R. HAMILTON, *supra* note 9 (§ 3.13(b) herein), at § 4.

19. Bean, *supra* note 1 (§ 3.11(a) herein), at 406.

20. *Id.*

21. 26 U.S.C.A. (I.R.C.) §§ 1361–1379 (West 1982 & Spec. Supp. 1982).

22. Subchapter S Revision Act of 1982 (1982 Act), Pub. L. No. 97-354, 96 Stat. 1669.

by subchapter S corporations and has made use of this business form a desirable one for drilling funds.

The number of shareholders permitted is now thirty-five,[23] and this number corresponds to the newly enacted exemptions under the federal securities laws.[24] Only individuals, with certain exceptions for estates and trusts, can be shareholders,[25] and there may be only one class of stock.[26] Under prior law, a subchapter S corporation was rarely selected for drilling funds because there was no pass-through to the shareholders of the depletions allowance.[27] The 1982 Act makes it possible for subchapter S corporation shareholders to compute cost and percentage depletion at the shareholder level in a manner similar to that used in the case of property held by a partnership.[28] Except in instances where a subchapter S corporation has previously been classified as a regular C corporation, passive investment income of 25 percent or more will not trigger termination of the corporation as it did previously.[29] Consideration of passive versus active income is not generally a problem for a drilling fund, because working interests, unlike royalties, are not considered passive.[30]

§ 3.14 Classification for Federal Income Tax Purposes

(a) General Considerations

If a limited partnership is classified as an association taxable as a corporation, no items of partnership gain, deduction, loss, or credit can flow through to the individual partners. Without this advantage, investors would be better off investing in shares of an established oil and gas company that has greater diversification.[31]

The criteria for determining if an organization is a corporation or a partnership were originally formulated in *Morrissey v. Commissioner*.[32] Corpo-

---

23. 26 U.S.C.A. (I.R.C.) § 1361(b)(1)(A) (West 1982 & Spec. Supp. 1982).

24. See *infra* § 5.23 for discussion of Regulation D, rules 505 and 506.

25. 26 U.S.C.A. (I.R.C.) §§ 1361(b)(1)(B), (c)(2), (c)(3) (West 1982 & Spec. Supp. 1982).

26. *Id.* § 1361(b)(1)(D).

27. See *infra* § 4.13 for discussion of depletion allowance.

28. 26 U.S.C.A. (I.R.C.) § 613A(c)(13) (West 1982 & Spec. Supp. 1982). Comparable partnership provision is in *id.* § 613A(c)(7)(D) (Supp. 1982).

29. *See id.* § 1362(d)(3) (1982 & Spec. Supp. 1982).

30. *See id.* § 1362(d)(3)(D).

31. Close, *Drilling Funds, The 1977 Perspective*, 28TH ANN. INST. ON OIL & GAS TAX 421, 549 (1977).

32. 296 U.S. 344, 56 S. Ct. 289, 80 L. Ed. 263 (1935).

rations and partnerships share two characteristics: associates and a joint profit motive. The four characteristics that may be used to distinguish a corporation from a partnership are:

1. continuity of life,
2. centralization of management,
3. liability for organization debts limited to organization property, and
4. free transferability of interest.[33]

In the 1950s, professional service organizations, such as doctors' groups, sought to be classified as corporations for tax purposes in order to qualify for pension and employee benefits that were not then available for limited partnerships. Following *United States v. Kintner*,[34] the IRS adopted the *Kintner* regulations in an effort to make it more difficult for professional associations to achieve corporate status.[35] The IRS adopted the "preponderance" test, which said that an organization must have a preponderance of the four *Morrissey* characteristics to be classified as a corporation; that is, more than half of the four characteristics had to be corporate, and if the characteristics were split evenly, the organization would be classified as a partnership.[36] To state it in another way, an organization has only to "fail" two of the four corporate criteria to gain partnership status.

With the enactment of professional corporation acts and their recognition for tax purposes, the issue of classification of professional service organizations dimmed in the years that followed *Kintner*. Nevertheless, the *Kintner* regulations, which made it difficult for an organization to be classified as a corporation, were still operative when the tax-shelter limited partnerships came into vogue. The IRS then wanted to tilt the regulations the other way so that these limited partnerships would more easily be classified as corporations and thereby lose their tax advantages. In 1975 the Tax Court held in *Larson*[37] that two real estate syndications, which seemed to meet the *Kintner* requirements, were associations taxable as corporations. The next day the Court of Claims decided a case on similar facts in exactly the opposite way.[38] The Tax Court then withdrew the *Larson* opinion and issued another opinion in the case, *Larson II*,[39] which returned to the *Kintner* standard, and held that a limited partnership that satisfied no more than two of the corporate charac-

---

33. *Id.* at 359.

34. 216 F.2d 418 (9th Cir. 1954).

35. Treas. Reg. §§ 301.7701-1, .7701-2, T.D. 7515, 42 Fed. Reg. 55,612 (1977).

36. *Id.*

37. Tax. Ct. Rep. (CCH) ¶ 33,471 (1975).

38. Zuckman v. United States, 524 F.2d 729 (Ct. Cl. 1975).

39. 66 T.C. 159 (1976).

teristics would not be classified as an association taxable as a corporation. Undaunted, the IRS proposed new regulations with new classification standards on January 5, 1977, which would have made it more difficult to obtain partnership classification. Treasury Secretary William Simon, however, withdrew these regulations on the same day that they were issued. Thus the *Kintner* regulations are still in force.

A drilling fund must be carefully organized so that it will not be taxed as a corporation. Following is a discussion of the four criteria.

(b) Continuity of Life

Generally in a partnership, dissolution may occur with death, insanity, or withdrawal of any member. Under section 21 of the Texas Uniform Limited Partnership Act,[40] which tracks the corresponding provision of the Uniform Limited Partnership Act,[41] the retirement, death, or insanity of a general partner dissolves the limited partnership unless the remaining general partners continue the business "(a) [u]nder a right so to do stated in the certificate, or (b) [w]ith the consent of all members." The Treasury regulations provide as follows: "If the retirement, death, or insanity of a general partner of a limited partnership causes a dissolution of the partnership, unless the remaining general partners agree to continue the partnership or unless all remaining members agree to continue the partnership, continuity of life does not exist."[42] In discussing the effect of local law in determining whether any member has the power of dissolution under a particular partnership agreement, the regulation further states that "a limited partnership subject to a statute corresponding to the Uniform Limited Partnership Act [lacks] continuity of life."[43]

In 1979 section 8 of the Texas Uniform Limited Partnership Act was amended to list certain powers whose possession or exercise by a limited partner would not constitute the partner's taking part in the control of the business.[44] The amended section includes a "partnership democracy" provision allowing the limited partners individually or by majority to approve certain matters, such as dissolution or sale of substantially all the assets of the partnership.[45] The appended comment of the committee of the State Bar of Texas that drafted the amendment states that, because of the fear of creating continuity of life as interpreted by the IRS, the list of powers in the amended

---

40. Tex. Rev. Civ. Stat. Ann. art. 6132a, § 21 (Vernon 1970).

41. Uniform Limited Partnership Act § 20.

42. Treas. Reg. § 301.7701-2(b)(1), T.D. 7515, 42 Fed. Reg. 55,612 (1977).

43. *Id.* § 301.7701-2(b)(3).

44. Tex. Rev. Civ. Stat. Ann. art. 6132a, § 8(b) (Vernon Supp. 1982–1983).

45. *Id.* § 8(b)(5).

section does *not* include "the rights of limited partners to remove and replace the general partner."[46] The committee warns that "before inserting removal rights, *coupled with* a right of less than all of the limited partners to replace the general partner and continue the partnership, it is wise to explore all the possible income tax consequences of this combination."[47]

If the partnership agreement is properly drawn, therefore, a drilling fund established as a limited partnership under Texas law should have little difficulty in failing the continuity-of-life test.

(c) Centralization of Management

A basic feature of a limited partnership is that management and control are exercised by the general partner and are thus centralized and that the limited liability of the limited partners will be destroyed if they exercise such control. The Code regulations, in considering the centralization issue, also make a distinction between limited partnerships in which the limiteds own "substantially all" the interest and those in which the general partners own a large share.[48] The IRS rule of thumb is that "substantially all" means 80 percent or more owned by the limited partners.[49] The distinction seems to be based on the premise that the general partner is acting more in a representative capacity if he owns little economic interest in the partnership and is therefore more analogous to a corporate officer.

Because of the nature of drilling funds, the limited partnership organized to drill for oil and gas will almost always be found to have centralization of management. The investors rarely have the requisite knowledge to participate meaningfully in the management of the fund, and the prospectus will often contain the language "The General Partner will conduct all Partnership activities . . ."[50] Furthermore, in many drilling funds the limited partners own 80 percent of the interests, particularly in the drilling stage of the program, which is the stage in which the sponsor exercises management almost exclusively. Therefore, limited partnership drilling funds will almost always pass the centralization-of-management test for corporate status because of control and percentage ownership.

---

46. *Id.* Comment at 27–28. The comment also states that the § 8(b) list is substantially like that in § 303(b) of the revised Uniform Limited Partnership Act (1976).

47. *Id.* (emphasis in original).

48. Treas. Reg. § 301.7701-2(c)(4), T.D. 7515, 42 Fed. Reg. 55,612 (1977).

49. Close, *supra* note 31 (§ 3.14(a) herein), at 438.

50. *See, e.g.*, Cola Petroleum, Ltd., 1979 Prospectus 17.

Concerning the percentage ownership of the general partner, it should be noted that, in order to obtain a private letter ruling from the IRS on the tax classification of a limited partnership, it is necessary that the general partner participate in every item of the partnership's income, gain, loss, deduction, or credit, at least to the extent of 1 percent.[51] Thus, drilling funds should always be structured to allocate no more than 99 percent of intangible drilling and development costs to the limited partners to preserve the opportunity of application for a private letter ruling.

(d) Limited Liability

A unique characteristic of a limited partnership is the combination of the personal liability of the general partner and limited liability of the limited partners. A corporation, on the other hand, has "no member who is personally liable for the debts of or claims against the organization."[52] Personal liability means that creditors of the organization may reach the personal assets of the member if the assets of the organization are insufficient.

In order to fail the limited liability test for corporate status, a limited partnership will have to firmly establish the liability of the general partner, because the regulations qualify the limited liability test if the general partner has no substantial assets that can be reached by a creditor.[53] Revenue Procedure 72-13 discusses the issue of requisite assets if a corporation acts as the sole general partner. To obtain an advance ruling from the Internal Revenue Service concerning tax classification, the revenue procedure states that the corporate general partner must meet three criteria: (1) no more than 20 percent of the stock of the corporate general partner may be owned by the limited partners; (2) the net worth of the corporate general partner must be 15 percent of the total assets of the limited partnership if the total assets are $2.5 million or less and 10 percent if the total assets are more than $2.5 million; and (3) if the corporate general partner is acting as the general partner of several limited partnerships, it must satisfy the net worth requirement for each limited partnership separately.[54]

Many practitioners recommend that a limited partnership with a sole corporate general partner adhere to these guidelines even though revenue procedures do not have the force of law.[55] One commentator, however, sug-

---

51. Rev. Proc. 74-17, 1974-1 C.B. 438.

52. Treas. Reg. § 301.7701-2(d)(1), T.D. 7515, 42 Fed. Reg. 55,612 (1977).

53. *Id.* § 301.7701-2(d)(2).

54. 1972-1 C.B. 735.

55. *But see* Humphrey, *supra* note 1 (§ 3.01 herein), at 11.

gests that, in his experience, many drilling funds have not sought a ruling on association status because of the stiff capitalization requirements of Revenue Procedure 72-13.[56] These funds have relied instead on legal opinions that the organization will be taxed as a partnership since the capitalization ratio is required only for a prior ruling and has not yet been litigated as an actual requirement for classification purposes.[57]

In any event, a limited partnership drilling fund should be able to fail the limited liability test for corporate status if structured with sufficient liability for the general partner. Coupling this result with a showing of lack of continuity of life will mean that the organization will be classified as a partnership for tax purposes, because it does not have two of the four corporate characteristics and therefore fails the preponderance test for corporate status cited above.[58]

Before 1979 the Texas Uniform Limited Partnership Act[59] provided in section 8 that a limited partner was not personally liable for the debts of the partnership *unless* he took part in the control of the business.[60] Section 8 caused substantial confusion about what activities would constitute "control." The confusion culminated in *Delaney v. Fidelity Lease Ltd.*,[61] where the court found limited partners personally liable when they acted as directors of the corporate general partner, because such action was considered control, even though creditors had not relied on the limited partners as general partners in extending credit.[62] The bar committee that drafted the amendment to section 8 in 1979 stated: "That decision [*Delaney*] created a windfall for careless or indifferent creditors at the expense of limited partners."[63]

Amended section 8 reverses the impact of *Delaney* by requiring that a creditor must reasonably rely on the limited partner as a general partner.[64] The section then provides a nonexclusive list of powers that may be exercised by a limited partner without such exercise being construed as control.[65] These powers include consulting with and advising the general partner, acting as a surety, acting as an officer or director of the corporate general

---

56. Bean, *supra* note 1 (§ 3.11(a) herein), at 402.

57. *Id.* at 402, 403.

58. *See supra* § 3.14(a).

59. Tex. Rev. Civ. Stat. Ann. art. 6132a (Vernon 1970 & Supp. 1982–1983).

60. 1955 Tex. Laws, ch. 133, § 8, at 472.

61. 526 S.W.2d 543 (Tex. 1975).

62. Attorney General John Hill later interpreted *Delaney* to mean that a corporation could not serve as general partner, despite statutory authority to the contrary. Op. Tex. Att'y Gen. No. H-1229 (Aug. 16, 1978). This opinion, coupled with *Delaney*, led to revision of the Limited Partnership Act.

63. Tex. Rev. Civ. Stat. Ann. art. 6132a, § 8 Comment at 27 (Vernon Supp. 1982–1983).

64. *Id.* § 8(a).

65. *Id.* § 8(b).

partner (reverses *Delaney*), and approving such material matters as dissolution, amendment of the partnership agreement, or sale of substantially all the assets.[66]

(e) Free Transferability of Interest

An advantage of ownership in a corporation is that, because such ownership is commonly in the form of stock that has a ready market, the interests in a corporation are freely transferable. Because shareholders cannot take part in the business of the corporation and are not personally liable, they are not selective in choosing one another. In contrast, a general partnership possesses mutual agency and unlimited liability of the partners, resulting in a careful selection of partners. A limited partnership resembles a corporation more closely in regard to the motivation for selecting partners, because of the passive nature of the limited partners and their limited liability. Thus free transferability of interest is desirable for corporations and limited partnerships alike, and it might initially appear that a limited partnership will have difficulty in distinguishing its organization from that of a corporation concerning this characteristic.

The Texas Uniform Limited Partnership Act provides that an assignee of a limited partnership interest is entitled only to a share of the profits and does not receive the other indicia of limited partnership ownership, which are the rights to request inspection of partnership books and to demand an accounting.[67] For an assignee to become a full-fledged, substituted limited partner, there must be approval by all the partners unless the agreement provides otherwise.[68] The Code regulations provide that free transferability, in the corporate sense, means that a "member must be able, without the consent of other members, to confer upon his substitute *all* the attributes of his interest in the organization."[69]

Thus, if the partnership agreement requires approval by the other members in order for an assignee to become a substituted limited partner, the limited partnership should fail the free-transferability test for corporate status. In *Larson II*, however, the Tax Court held that approval by the general

---

66. The revised Uniform Limited Partnership Act (1976) expressly permits a corporation to act as a general or limited partner (§ 101) as well as permitting limited partners to be employees, and presumably officers also, of the corporate general partner (§ 303(b)(1)). The revised Act has been adopted by Wyoming, Connecticut, and Minnesota, and other states are expected to follow suit.

67. TEX. REV. CIV. STAT. ANN. art. 6132a, § 20(c) (Vernon 1970).

68. *Id.* § 20(d).

69. Treas. Reg. § 301.7701-2(e)(1), T.D. 7515, 42 Fed. Reg. 55,612 (1977) (emphasis added).

partner, if that approval cannot reasonably be withheld, is not a sufficient barrier to transferability.[70] One preliminary prospectus reviewed, anticipating this problem, provides that the general partner may arbitrarily and capriciously withhold consent to transfer.[71] Most of the drilling fund prospectuses and offering memoranda reviewed provide only for the general partner's approval and hence under relevant statutory and case law would not fail the free-transferability test for corporate status. If the offering has already failed to have two corporate characteristics (continuity of life and limited liability), however, it will have qualified as a limited partnership notwithstanding free transferability of interest.

Further limits on transferability may be provided by the relevant securities laws. These restrictions are discussed in chapter 5 of this book. Although the IRS regulations and the commentators speak only of restrictions on transferability in terms of consent by other members, it appears that the limitations imposed by the Securities Act of 1933 in regard to transfer of restricted shares[72] can provide the requisite limitation.

(f) Summary

In summary, a drilling fund established as a limited partnership in order to be classified as a partnership for tax purposes should possess no more than two of the four corporate characteristics originally set forth in *Morrissey*.[73] Continuity of life and limited liability are the two criteria by which a limited partnership can most easily distinguish itself from a corporation. Lack of free transferability is easily achieved but seems to be less desirable than meeting the continuity and limited liability tests. Centralization of management is almost always present in a drilling fund, and hence a limited partnership is not distinguishable from a corporation in this respect. The drilling funds reviewed vary in their approach to the classification problem. Some rely on attorneys' opinions, and others seek a prior letter ruling from the IRS. If an attorney's opinion is relied on, it should state that it represents only a legal judgment and is not binding on the IRS. However, if a letter ruling is sought from the IRS, the organization of the limited partnership must meet the requirements set forth in Revenue Procedures 74-17 and 72-13, which are discussed at *supra* section 3.14(c) and (d).

---

70. Larson II, 66 T.C. 159, 183 (1976).

71. Sterling Gas Drilling Fund, Preliminary Prospectus, Exhibit A, Limited Partnership Agreement § 9.4 (Aug. 1978).

72. See discussion in *infra* ch. 5.

73. Morrissey v. Commissioner, 296 U.S. 344, 56 S. Ct. 289, 80 L. Ed. 263 (1935).

Finally, the Code regulations provide that in addition to the four factors discussed, other factors may be taken into account.[74] A request was recently made to the IRS for a ruling on what constituted "other factors," and a fairly convoluted answer was provided.[75] The ruling says that the IRS will follow *Larson II* and that *Larson II* found the following factors *either* part of one of the four major factors *or* of no critical importance (the ruling gives no clue which is which):

1. division of limited partnership interests into units or shares and marketing like securities,
2. the right of the managing partner to retain or distribute profits,
3. the right of the limited partners to vote to remove general partners and to sell substantially all the assets,
4. interest in the limited partnership represented by certificates,
5. observance of corporate formalities,
6. requirement of signatures for the partnership agreement, and
7. pooling of investments while limiting liability.[76]

The ruling concludes by stating that the IRS will not consider the above factors except as they bear on the four major characteristics that have been discussed. The second factor listed appears to bear on centralization of management, and, since a limited partnership drilling fund cannot be differentiated from a taxable association on this factor, the ruling seems to have no bearing. The third factor, the right of removing a general partner or selling assets, bears on continuity of life and can be handled to avoid continuity of life in a carefully drawn partnership agreement.[77] The other five criteria appear to be of no critical importance and therefore should have little bearing on the four major factors.

§ 3.15 Compensation to Sponsor

(a) General Considerations

An investor will want to analyze the sponsor's compensation carefully. This compensation usually comes from two sources: the initial proceeds of the drilling fund or participation in drilling fund revenues. The source and con-

---

74. Treas. Reg. § 301.7701-2(a)(1), T.D. 7515, 42 Fed. Reg. 55,612 (1977).

75. Rev. Rul. 79-106, 1979-1 C.B. 448.

76. *Id.*

77. *See supra* § 3.14(b).

figuration of the compensation need to be examined relative to the sponsor's participation in the costs of the fund.

(b) Sources of Compensation

The sponsor will often receive a one-time management fee from the proceeds. If the fund is not registered with the SEC, the fee may include the costs of selling the offering and should be contrasted with the brokerage costs incident to a public offering, which can run as much as 8.5 percent. Organization costs, in addition to a management fee, are sometimes charged to the fund and may include executive salaries of the sponsors. Further, the sponsor may retain an interest in the revenues of the partnership or venture as compensation. This usually takes one of two forms. The sponsor may retain an overriding royalty, which can be converted to a working interest when the investors have recouped their costs. Because the override is paid from production revenues, if no oil or gas is discovered, no compensation to the sponsor results from this arrangement. The sponsor alternatively may choose to retain a net profits interest, which differs from an override in that the former is paid as a percentage of profit *after* all production expenses are paid. Typically the net profits interest arrangement might work like this: of the gross proceeds, the landowner or mineral royalty share is 20 percent and production costs are another 30 percent, leaving 50 percent of the gross proceeds for the investors. If a 15 percent net profits interest is retained, the sponsor will receive 15 percent of 50 percent (7.5 percent) as compensation. If an overriding royalty is retained, however, it applies to the gross proceeds in the same manner as the landowner or mineral royalty.

(c) Participation in Costs

Another factor to consider is the sponsor's participation in the costs. Does the sponsor pay any costs of the first well or does he get a "free look" and bear no costs if the well is dry? Is the sponsor contributing leases? These considerations operate in conjunction with the relative risks of the program—that is, the degree to which the wells to be drilled are exploratory or developmental.

Various guidelines are given, but because the sponsor's compensation can take so many forms it is difficult to apply a rule of thumb. An investment advisor who was consulted said that he looked for a fund structured so that, if the limited partners lose money because of a dry hole, the general partner will lose money as well; that is, the sponsor participates in the costs of the first well. Drilling funds of the particular type that this investment advisor recommends are 75 percent developmental and 25 percent exploratory and thus have a high probability of achieving a return of the initial investment. However, if a sponsor is organizing a fund to drill predominantly exploratory wells, the sponsor's chances of recouping first-well costs are dramatically decreased. It is difficult to see how a sponsor can afford to share first-well costs in any meaningful way on repeated exploratory wells. An investor should bear in mind that a sponsor can be involved in a drilling fund in two very distinct ways: as a business arrangement selling skills in the oil business or as an investor. If the risks of the venture are very high, the sponsor may choose to participate primarily in the former way and receive a larger part of the compensation in the form of a management fee with little participation in costs.

Even if the sponsor is obliged to participate in costs, a close reading of the prospectus may reveal that he is not obligated to put up his money for several years and that, in fact, his balance sheet does not support his ability to pay his share of costs from the outset. In this situation the sponsor may never be out of pocket and yet may receive 40 percent or more of the revenue produced by the oil and gas discovered.[78] A Pennsylvania oil company, whose president cites a need for better investor return, offers a 120 percent payout to investors before the sponsor takes anything and thereafter shares in a 25:75 percent ratio.[79]

(d) Guidelines

The Texas Securities Board, in its guidelines for registration of drilling funds, sets a 15 percent maximum for the dollar amount paid to the sponsor from the fund proceeds from which organization and offering expenses, as well as the management fee, must be paid.[80] Further participation must be "reasonable" and may include retained interests, such as an overriding royalty or net profits interest, that are "generally accepted as reasonable in the industry and are justified."[81] One practitioner recommends a ceiling of 10–12 percent for the management and brokerage fees, including 8.5 percent for the brokerage fee and 3 percent for overhead.[82] However, he also recommends that if the sponsor's interest in the program will be as high as 40–50 percent

78. Metz, *Small Investors Still Like Oil, Gas Programs*, Houston Chronicle, Aug. 21, 1980, at 10, sec. 2.

79. *Id.*

80. State Sec. Bd., 7 Tex. Admin. Code § 121.2(d) (Shepard's, Sept. 1, 1981).

81. *Id.*

82. Humphrey, *supra* note 1 (§ 3.01 herein), at 6, 7.

there should be a mandatory contribution even if all the initial prospects are dry holes.

The prospectuses reviewed have varied enormously in terms of compensation. The outlines of several are sketched below to describe the variation:

Drilling Fund A

1. Selling commission of 8 percent.
2. Selling commission and organization costs financed by a bank loan paid 99 percent by limited partners from first revenues. If there are no revenues, recourse is to the general partner in his individual capacity.
3. Overhead expenses, which include executive salaries, paid 99 percent by limited partners. These expenses will be no more than 6 percent of the subscription amount.
4. Share of 10 percent by general partner of all costs and revenues.
5. The sponsor or general partner retains a 15 percent net profits interest.
6. Proposed activity: 75 percent developmental, 25 percent exploratory.

Drilling Fund B

1. Selling commission of 7.5 percent.
2. Selling commission and organization costs financed by a bank loan paid 99 percent by limited partners from first net revenues. If there are no revenues, recourse is to the general partner in his individual capacity.
3. Management fee of 5 percent of the initial partnership capital.
4. Overhead expenses are paid 70 percent by limited partners, and these expenses do not include executive salaries. Maximum of 6.5 percent of subscription proceeds.
5. Share of 5 percent of all costs to casing point, share of 30 percent of costs thereafter, 30 percent participation in all revenues.
6. Proposed activity: 75 percent developmental, 25 percent exploratory.

Drilling Fund C

1. No selling or organization costs.
2. Management fee of 15 percent of initial partnership capital, which includes overhead for first and subsequent years.
3. No share of costs of drilling or completing the first well; thereafter 25 percent of costs.

4. Overriding royalty interest of 12.5 percent on net proceeds (identical to net profits interest) during payout with option to convert to 25 percent working interest after recoupment.
5. Proposed activity: 100 percent exploratory.

§ 3.16 Analysis of Risk

In a fund where the leases to be drilled are preselected, one practitioner recommends application of geologic and reservoir engineering to the particular prospects to be drilled, combined with statistical probability analysis.[83] The resulting predictions applied to a single prospect will, statistically speaking, be wrong but applied to a hundred prospects may be useful. This practitioner opined that a broker should have such an analysis prepared as part of its due diligence review, but an investment counselor who was interviewed was unfamiliar with this technique and disparaged its use. Because an investor rarely looks at a hundred drilling funds, the marginal utility of such an analysis for an investor is questionable; however, a sponsor evaluating many properties and establishing many funds may find it useful.

The first step in this type of analysis is evaluation of each prospect according to its geologic information. Second, production statistics of other wells in the reservoir are studied. A risk factor is assigned to the prospect, such as .6 if it is predicted that the projected reserves and economics will be realized 60 percent of the time. Finally, an analysis of probability is done for the 50 percent income tax bracket. The resulting study indicates expected profit, expected return of the initial risk investment, and the probability of such return. The accepted risk recommended by Humphrey is a probable two-for-one return on investment for low-risk programs, three-for-one for balanced programs, and four-for-one for high-risk programs.[84] This particular analytical device is unique, and its relative merit should be carefully evaluated.

§ 3.17 Leverage

With the exception of holding real property (other than mineral property) and equipment leasing by closely held corporations,[85] an investor may de-

---

83. *Id.* at 10.
84. *Id.* at tab 7.
85. 26 U.S.C.A. (I.R.C.) § 465(c)(3), (4) (West Supp. 1982).

duct losses and expenses only up to the amount for which he is "at risk."[86] The amount at risk with respect to an oil and gas property equals the amount of money invested and the adjusted basis of other property contributed to the drilling fund, plus amounts borrowed from unrelated lenders for which the investor is personally liable.[87] Section 703[88] applies the "at risk" rules to partnerships, but with the advantage that the rules apply to the aggregate partnership properties rather than a "single activity,"[89] thus permitting partners to offset losses from one property within the partnership with income from another.[90] Before the "at risk" rules were adopted in 1976, the adjusted basis of a partner's share in a drilling fund could be increased by the limited partner's share of nonrecourse loans to the partnership. However, the IRS viewed this action as an enormous abuse, and in their training manual of that year warned agents that "it is extremely difficult, if not impossible, to secure a non-recourse loan, secured only by undeveloped [oil and gas] properties, from a responsible lender."[91] The "at risk" rules passed in 1976 specified that nonrecourse financing could not be added to the adjusted basis[92] in computing losses.

It is, in fact, difficult to obtain nonrecourse loans to finance exploratory drilling. However, one practitioner suggests that increased flexibility in investment planning can be obtained by financing 70–80 percent of the investor's commitment with letters of credit.[93] Such letters of credit have full recourse and therefore add to the adjusted basis. By prearrangement with a bank, a partnership can receive a loan equal to the pool of the letters of credit provided. If commercial production is obtained by the drilling fund, revenues will begin about the time that interest is due on the loan, and the interest payments are deductible as well. The investor is thus allowed to maintain a more liquid portfolio. Another advantage may be that the investor can hold an investment that normally might have to be sold in order to meet the drilling fund investment commitment. One disadvantage is that, because of the interest payments, an unsuccessful drilling program will result in larger losses under the letter-of-credit arrangement than a cash investment would. Further, this method of financing might jeopardize an investor's line of credit for other purposes.

---

86. *Id.* § 465(a).

87. *Id.* § 465(b) (1978 & Supp. 1982).

88. *Id.* § 703 (1982).

89. *See id.* § 465(c)(2) (1978 & Supp. 1982).

90. Close, *supra* note 31 (§ 3.14(a) herein), at 492.

91. IRS Tax Shelter Training Manual § 3.201 (1976).

92. 26 U.S.C.A. (I.R.C.) § 465(b)(2)(A) (West 1978).

93. Humphrey, *supra* note 1 (§ 3.01 herein), at 7–10.

§ 3.18 Evaluation of Sponsor

This section should be printed in red to emphasize its importance. The critical factor that distinguishes a drilling fund from a real estate syndicate or other tax-shelter investment is that it is impossible to provide enough surveillance of oil and gas operations to ensure a fair deal. The complexity of the business, the degree of risk, and the largely unknown operation of the venture make the oil and gas business particularly susceptible to unethical and even fraudulent promoters.

Every person interviewed began by emphasizing that an investor must *KNOW THE SPONSOR!* If the investor does not personally know the sponsor, he may rely on a trusted broker as a substitute. Little else can provide the requisite safeguards. An investor who personally knows the sponsor is aware of the sponsor's background in the oil and gas business and his reputation in the business community.

Evaluation of the sponsor's business background and reputation from impersonal sources is no substitute for personal knowledge. Thus, fraudulent oil and gas schemes historically have involved out-of-state investors who have no opportunity to know the sponsor. The schedule D scandals that were so prominent in 1976 in Texas are discussed at *infra* section 5.28. These scandals highlight the potentials for fraudulent sponsors.

CHAPTER FOUR

Tax Aspects

[*Unassigned section numbers are reserved for possible use in supplementation.*]

I. Special Allocations

§ 4.01 Significance of Expense Allocation

The single most critical tax aspect in the decision to invest in a drilling fund is the ability to deduct the intangible drilling costs (IDCs) as an expense when paid[1] rather than to capitalize them and recover them through cost depletion. Because IDCs are basically the costs of drilling the well,[2] they occur in the initial stages of the partnership or venture and hence can provide an early deductible expense. This attractiveness to potential investors is further enhanced by the ability to prepay IDCs if certain conditions are met.[3]

§ 4.02 Allocation in the Carried–Interest Arrangement

In the classic carried-interest oil deal that is not organized as a partnership, the owner of the lease assigns a percentage of the working interest in the lease to the carrying party in return for that party's agreeing to drill the well. The parties agree to share production revenue equally or in some predetermined percentage. In this case section 1.612-4(a) of the Treasury Regulations[4] provides that the parties may deduct IDCs only to the extent of their fractional working interests. Thus the carrying party, who pays all drilling costs, may only deduct a portion of the IDCs if he does not own the entire working interest. This deduction is often not a sufficient tax incentive to encourage investment by the carrying party. Thus, to circumvent the regulation, sponsors began assigning 100 percent of the working interest during the complete payout period to the carrying party with a reversion of a per-

1. 26 U.S.C.A. (I.R.C.) § 263(c) (West Supp. 1982), *as amended by* Tax Equity and Fiscal Responsibility Act of 1982; Treas. Reg. § 1.612-4, T.D. 6836, 30 Fed. Reg. 8902 (1965). The Tax Equity and Fiscal Responsibility Act of 1982, Pub. L. No. 97-248, 96 Stat. 324, reduces the amount deductible by corporations (other than subchapter S corporations) for IDCs by 15%. *Id.* §§ 204(c)(1), 204(a) (to be codified at 26 U.S.C.A. §§ 263(c), 291). The same Act also allows limited partners the election of a 10-year amortization of IDCs and other individuals the election of 5-year depreciation. *Id.* §§ 204(c)(1), 201(c)(1) (to be codified at 26 U.S.C.A. §§ 263(c), 58(i)).

2. IDCs are those costs of drilling and development that produce no asset with salvageable value. Treas. Reg. § 1.612-4, T.D. 6836, 30 Fed. Reg. 8902 (1965).

3. *See* Rev. Rul. 71-252, 1971-1 C.B. 146; *id.* 71-579, 1971-2 C.B. 225; Cheroff v. Commissioner, 40 T.C.M. 183 (Apr. 21, 1980). *But see* Rev. Rul. 80-71, 1980-1 C.B. 7.

4. T.D. 6836, 30 Fed. Reg. 8902 (1965).

centage interest to the carried party at that time. The IRS supports deduction of all the IDCs by the carrying party in this situation.[5]

§ 4.03 Allocation in the Limited Partnership

(a) Reasons To Use Limited Partnership

In the carried-interest situation discussed above, sponsors were dissatisfied not to participate in production revenues until after the complete payout period and thus seized on the limited partnership vehicle as a means of sharing in revenues at an earlier point while still satisfying the investor's requirements for deduction of all the IDCs. A partnership owns all leases acquired in the name of the partnership and, as an incident of ownership, the corresponding working interest. Therefore, the partnership is entitled to the entire deduction of IDCs. The partnership may then allocate this deduction to the various partners in any way it chooses as provided for in the partnership agreement[6] if the allocation has substantial economic effect.[7]

(b) Earlier IRS Treatment

The regulations, in defining "substantial economic effect," state that an allocation will not be considered to have substantial economic effect if the principal reason for the allocation is avoidance or evasion of federal income tax.[8] The regulations continue with relevant circumstances to be considered:

1. whether the allocation has a business purpose,
2. whether the allocation has a substantial economic effect that actually affects the dollar amount of the partners' shares of the total partnership's income or loss independently of tax consequences,
3. whether related items are allocated in the same way,
4. whether the allocation was made without recognition of normal business factors and only when the amount of the item could be estimated,
5. the duration of the allocation, and
6. the overall tax consequences.[9]

How can one define the substantial economic effect required by the Internal Revenue Code by considering "substantial economic effect" (item 2

---

5. Rev. Rul. 71-207, 1971-1 C.B. 160; *id.* 69-332, 1969-1 C.B. 87.

6. 26 U.S.C.A. (I.R.C.) § 704(a) (West 1982).

7. *Id.* § 704(b)(2).

8. Treas. Reg. § 1.701-1(b)(2), T.D. 6771, 29 Fed. Reg. 15,571 (1964).

9. *Id.*

above) as a relevant circumstance? The reasoning seems circular. It appears that this circularity resulted from an amendment to the Code that elevated substantial economic effect from a regulatory factor to a Code requirement with no subsequent amendment of the regulations for clarification.

Equally puzzling is the only revenue ruling on special allocations in an oil and gas partnership.[10] The example posed is of two partners contributing all the drilling costs and taking all the IDC deductions and a third party contributing the lease. Partnership income is allocated 3/32, 3/32, and 26/32, respectively, from the outset; presumably proceeds from the sale or liquidation of the partnership would be similarly divided.[11] In Revenue Ruling 68-139,[12] the IRS approved this allocation "unless examination discloses that the principal purpose of the allocation is the avoidance or evasion of Federal income tax." This last stipulation, in the opinion of many commentators, renders the ruling unreliable,[13] if not "devoid of precedential value."[14]

Despite the deficiencies in reasoning in Revenue Ruling 68-139, most tax advisors interpret it to mean that within a limited partnership the general partner *may* share in revenues before complete payout to the limited partners.[15] Their interpretation is buttressed by an example given in the regulations.[16] In this example one partner contributes $25,000 plus services and the other contributes $100,000 in return for certain deductions, which are disproportionate to the amount of investment. Income is allocated 10 and 90 percent respectively until payout, when the allocation becomes 50 and 50 percent. The IRS indicates this allocation will be recognized but again throws in the caveat that such recognition is "in the absence of other circumstances showing that its principal purpose was tax avoidance or evasion."[17] Tax advice thus supported disproportionate allocation of IDCs within a limited partnership, and such allocation proceeded without incident in the oil and gas business until recently.

(c) Emerging Emphasis on Capital Accounts

One case, *Orrisch v. Commissioner*,[18] was prescient but caused few ripples in oil and gas limited partnerships at the time it was decided. A brief discus-

---

10. Rev. Rul. 68-139, 1968-1 C.B. 311.

11. DeBerry, *Organizing and Terminating the Oil and Gas Partnerships; Selected Current Problems*, 31ST ANN. INST. FOR OIL & GAS LAW & TAX. 408 (1980).

12. 1968-1 C.B. 311.

13. Casey, *Allocations in Oil and Gas Partnerships*, 29TH ANN. INST. FOR OIL & GAS LAW & TAX. 525, 536 (1978).

14. DeBerry, *supra* note 11, at 408.

15. *Id.*

16. Treas. Reg. § 1.704-1(b)(2), example 5, T.D. 6771, 29 Fed. Reg. 15,521 (1964).

17. *Id.*

18. 55 T.C. 395 (1970), *aff'd per curiam*, 31 A.F.T.R.2d 73 (9th Cir. 1973).

sion of capital accounts will facilitate an understanding of *Orrisch* and subsequent cases and rulings.

A partner's capital account is "simply a measure of a partner's entitlement, from time to time, to the proceeds of liquidation of the partnership."[19] If *L* contributes the lease worth $10x$ and *D* contributes $100x$ for drilling expenses and deducts the IDCs, the capital account should be as shown in table 4.1. If the partnership is sold at this point, the taxable gain is the selling price less the cost of the lease. Questions arise on how to divide the taxable gain and how to divide the cash proceeds. If taxable gains are allocated so that the capital accounts will be equalized, *D* will have to report a larger gain for tax purposes, but the proceeds can be distributed equally (see table 4.2). The foregoing hypothetical distribution applies if the proceeds of sale are sufficient to provide enough taxable gain to equalize the capital accounts. Consider the different situation presented in table 4.3. In the situation presented in table 4.3, the *Orrisch* interpretation is that *D* has received a disproportionate benefit from the deduction of IDCs, because *D* shares equally in the proceeds of the sale and is not required to equalize the capital accounts. If *D* had deducted only $90x$ of IDCs, *D*'s capital account would have equaled *L*'s and there would be no imbalance. The cases that have considered special allocations have focused on this imbalance, and, while the cases have generally concerned real estate tax shelters and special allocations of depreciation, the principles are the same.

TABLE 4.1

CAPITAL-ACCOUNT ADJUSTMENT FOR IDC DEDUCTION

	L	*D*
Original contribution	$10x$	$100x$
Deduction for IDCs	-0	$-100x$
Adjusted capital account	$10x$	0

19. DeBerry, *supra* note 11 (§ 4.03(b) herein), at 410.

TABLE 4.2

CAPITAL-ACCOUNT ADJUSTMENT AFTER HYPOTHETICAL SALE OF 100x

Note: Assume proceeds of sale are 100x and taxable gain is 90x (100x less lease cost of 10x).

	L	*D*
Capital account from table 4.1	10x	0
Allocation of gain	40x	50x
Adjusted capital account	50x	50x
Distribution of proceeds	50x	50x

TABLE 4.3

CAPITAL-ACCOUNT ADJUSTMENT AFTER HYPOTHETICAL SALE OF 10x

Note: Assume proceeds of sale are 10x and taxable gain is 0 (10x less lease cost of 10x).

	L	*D*
Capital account from table 4.1	10x	0
Allocation of gain	0	0
Adjusted capital account	10x	0
Distribution of proceeds	5x	5x

In *Orrisch* the taxpayer was specially allocated items of depreciation that created an imbalance in the capital accounts, and it was agreed that if the property were sold, the taxpayer's account would be charged with taxable gain to equalize the capital accounts (see table 4.4). Consider two hypo-

TABLE 4.4

Orrisch CAPITAL-ACCOUNT ADJUSTMENT FOR DEPRECIATION

	Taxpayer	Partner
Original contribution	$27,000	$12,000
Depreciation	−52,000	−12,000
Adjusted capital account	($25,000)	$ 0

thetical sales and their impact on the capital accounts, as reflected in table 4.5 The first hypothetical situation of $100,000 gain is analogous to the situation the partners agreed to in *Orrisch*, but the Tax Court held that the result is merely a "trade of tax consequences"[20] and has no substantial economic effect, because the taxpayer did not agree to bear the economic burden should the sale produce a taxable gain that is not enough to balance the capital accounts, as in the second hypothetical sale. If the sale is for $10,000 gain, as in the second situation, presumably, if the taxpayer agrees in the partnership agreement to give the entire $10,000 in cash to his partner plus enough cash ($2,500) to equalize the capital accounts, the burden will be considered borne.

This focus on capital accounts, which began in *Orrisch*, reappeared in the *General Explanation of the Tax Reform Act of 1976*:

> The determination of whether an allocation may actually affect the dollar amount of the partner's shares of total partnership income or loss independent of tax consequences will to a substantial extent involve an examination of how these allocations are treated in the partner's capital account for financial (as opposed to tax) accounting purposes; This assumes that these accounts actually reflect the dollar amount that the partners would have the right to receive upon liquidation of the partnership.[21]

20. Orrisch v. Commissioner, 55 T.C. 395, 403 (1970), *aff'd per curiam*, 31 A.F.T.R.2d 73 (9th Cir. 1973).

21. JOINT COMM. ON TAX., 94TH CONG., 2D SESS., at 95 n.6 (1976).

TABLE 4.5

Orrisch Capital-Account Adjustment
for Two Hypothetical Sales

	Taxpayer	Partner
Capital account from table 4.4	($25,000)	$ 0
1. Gain from sale is $100,000		
Allocation of gain	62,500	37,500
Adjusted capital account	37,500	37,500
Distribution of proceeds	50,000	50,000
2. Gain from sale is $10,000		
Allocation of gain	10,000	0
Adjusted capital account	(15,000)	0
Distribution of proceeds	$ 5,000	$ 5,000

(d) Effect of Failure To Maintain Capital Accounts

Since 1976 many tax practitioners have been concerned that, even if there is full recoupment of costs for the limited partners, a special allocation may be disallowed if capital accounts are not maintained.[22] This concern was reinforced by three subsequent cases disallowing special allocations, though in each of these cases the omission seems more egregious than in *Orrisch* because in these cases there is not even a provision for chargeback to the capital accounts if the taxable gain is sufficient, much less if it is insufficient.[23]

A letter ruling from 1977, which was made public in March 1979 through the Freedom of Information Act, disallowed a special allocation in an oil and gas limited partnership on these facts: *G* and *L* shared costs of

22. Bean, *supra* note 1 (§ 3.11(a) herein), at 389.

23. Boynton v. Commissioner, Tax Ct. Rep. Dec. (P-H) ¶ 72.95 (1979); Holladay v. Commissioner, Tax Ct. Rep. Dec. (P-H) ¶ 72.51 (1979); Magaziner v. Commissioner, Tax Ct. Mem. Dec. (P-H) ¶ 78,205 (1978).

drilling and deduction of IDCs in a 5:95 percent ratio, and income was also allocated 5 and 95 percent until *L* recouped the initial contribution, whereupon the sharing ratio became 40:60 percent for income and assets on liquidation.[24] If liquidation took place before payout, *G* would receive 40 percent of the partnership assets, but his interest was subject to the right of *L* to recover a share of his investment through a net profits inerest. While this ruling was being appealed, several other letter rulings concerning special allocations in oil and gas partnerships were made.

A 1980 letter ruling also disallowed a special allocation of IDCs to certain partners to the extent of their cash contributions because all the nondeductible equipment expenditures plus some of the IDC expenditures were allocated to another partner from his cash contribution.[25] The IRS, citing Revenue Ruling 68-139 and example 5 of Treasury Regulation section 1.704-1(b)(2), concluded that there was no business reason for certain partners' cash contributions to be allocated only to IDCs and held that the partners should be allocated their pro rata shares of the nondeductible cash expenditures. The flaw in this partnership agreement seems to be that the partner who was allocated the equipment expenses and some of the IDCs contributed a much larger percentage of cash to the venture than the partners receiving most of the IDC deductions, and the IRS found no "business reason" for the special allocation. Furthermore, the payout obligation to the partners was not tied to the allocation of deductions.

The next year a partnership special allocation was disapproved on the following facts: *A* provided all the drilling costs in return for a 50 percent ownership of the wells, and *A* received all the IDC deductions.[26] No provision was made for allocation of income, gain, or loss, and, lacking such a partnership provision, the distribution is made based on each partner's interest in the partnership, or 50 percent in this case.[27] Because the partnership agreement had no provision requiring partners to consider capital accounts on liquidation, the IRS reasoned that allocation to *A* of 100 percent of the IDCs did not affect the dollar amount that *A* would be entitled to on liquidation and thus lacked substantial economic effect.

Almost immediately after this ruling, the IRS approved a special allocation of IDCs in a situation in which the allocation reduced the capital ac-

---

24. Letter Rul. 7707260880A (Technical Advice Memorandum), 9 FED. TAXES (P-H) ¶ 55,192 (1979).

25. Letter Rul. 8008054 (Technical Advice Memorandum, Nov. 28, 1979), LETTER RUL. REP. (CCH) Rep. No. 15, Mar. 7, 1980.

26. Letter Rul. 8133028 (Technical Advice Memorandum) (CCH), May 19, 1981.

27. 26 U.S.C.A. (I.R.C.) § 704(b)(1) (West 1982).

count of the partner receiving the deduction and thereby reduced his claim to the partnership assets on liquidation even though there was no specific requirement that the capital account be balanced.[28]

(e) Court Consideration of Special Allocations

The letter rulings discussed above appear inconsistent in observing the *Orrisch* principles of maintenance of capital accounts that are given effect in liquidation. The 1977 letter ruling cited earlier as denying a special allocation of IDCs was based on facts utilizing a net profits interest to balance capital accounts in a "pre-payout" liquidation.[29] This inconsistency was removed by *Hamilton v. United States*,[30] which reversed the 1977 letter ruling despite novel loan theories advanced by the IRS.

One case seems something of a maverick and deserves consideration.[31] In *Allison* the Court of Claims never mentions capital accounts but, rather, emphasizes the necessity of special allocation of IDCs as an inducement to encourage investment in high-risk, wildcat drilling programs. This case involved exploratory drilling in a thirty-two-million-acre area of submerged land off the coast of Sumatra, and the court said that special allocation provisions were customary in the oil and gas business and were indeed necessary if investors were to participate in such risky ventures. Perhaps a different standard is to apply for wildcat drilling.

(f) Guidelines for Drafting

The result of the foregoing decisions and rulings is that commentators and practitioners alike are viewing limited partnerships as less-favorable vehicles than the carried-interest arrangement if capital accounts must be maintained and given effect in liquidations.[32] For a limited partnership, Casey proposes the following drafting suggestions:

1. Capital accounts should be established.
2. Items of net income, net loss, deduction, and credit should be defined as they are determined for federal income tax purposes.

---

28. Letter Rul. 8133021 (Technical Advice Memorandum) (CCH), Apr. 30, 1981.

29. *See* Letter Rul. 7707260880A, *supra* note 24.

30. 82-2 U.S. Tax Cas. (CCH) ¶ 9546 (Ct. Cl. 1982).

31. Allison v. United States, 82-1 U.S. Tax Cas. (CCH) ¶ 9163 (Ct. Cl. 1982).

32. DeBerry, *supra* note 11 (§ 4.03(b) herein), at 421; Bean, *supra* note 1 (§ 3.11(a) herein), at 390; SHEPHERD & HICKS, 1980 OIL AND GAS PROGRAM, DESCRIPTIVE MEMORANDUM 9.

3. The partnership agreement should provide that the above items are charged to the partners' capital accounts.
4. The partnership agreement should provide that, upon liquidation, proceeds will be distributed according to the partners' proportionate capital account balances.
5. To satisfy the *Orrisch* requirement that partners agree to eliminate deficit balances, there should be an unrecorded agreement to this effect. It should be unrecorded so that limited partners do not become exposed as general partners to the creditors of the partnership.[33]

Although this last requirement might seem to impose an undue burden on the limited partners, deficits in capital accounts are unlikely in drilling funds because they are not leveraged. Thus, the eventuality of having to restore a deficit is remote.

§ 4.04 Alternatives to Use of Limited Partnerships

(a) Introduction

Two approaches have been recommended recently to avoid the problems posed for special allocations in limited partnerships, and both are progeny of the classic carried-interest arrangement. These approaches are the "tax partnership" arrangement and the joint venture form, discussed in the next two subsections.

(b) The Tax Partnership

One approach suggested to avoid the special allocation problems faced by limited partnerships is a "tax partnership" in which there are coowners of the lease(s) who have *agreed not to be excluded from the application of subchapter K.*[34] Historically, the difficulty in distinguishing between coowners and partners for tax purposes has arisen in the context of oil and gas coowners who agree to share expenses for the joint exploration, development, and production of their property.[35] The Uniform Partnership Act's definition of partnership, which is common to all states, essentially includes joint ventures; the Act provides that "[a] partnership is an association of two or

33. Casey, *supra* note 13 (§ 4.03(b) herein), at 541–43.

34. DeBerry, *supra* note 11 (§ 4.03(b) herein), at 422 (emphasis in original).

35. Chapoton, The Subchapter K Election: What, When, Why, and How, at 2 (unpublished manuscript, made available by Mr. Ed Turley).

more persons to carry on as co-owners a business for profit."[36] The distinction between a partnership and a joint venture lies only in the scope of activity, inasmuch as a joint venture usually is an agreement covering a single project. Uncertainty exists, however, in distinguishing between coownership on the one hand and a partnership or a joint venture to explore for oil and gas on the other. Coownership problems are further compounded in Texas because either coowner can drill on the land and, while net revenue must be shared with other coowners, losses must be borne alone if the other coowners have not consented to the drilling. This situation creates the need for some sort of operating agreement between coowners, which in turn implies a business and runs afoul of the regulations because a coowner is thus classified (as opposed to a joint venture or partnership) only if the coowners do not actively carry on a trade or business.[37]

In 1953 the Tax Court resolved the definitional question by holding that coownership with a joint operating agreement creates a partnership for tax purposes.[38] The following year the amended Code added a new section granting certain oil and gas ventures the right to be excluded from the partnership rules of the Code.[39] There are three requirements for this election: (1) the members must be able to compute their income separately, (2) they must reserve the right to take the oil and gas in kind, and (3) they must not jointly sell the oil and gas for a period of more than one year.[40]

If a coownership of an oil and gas property with an operating agreement fails to meet these requirements for election out of subchapter K because of joint sales, it will probably be considered a partnership for tax purposes. It must be noted, however, that such a coownership can come perilously close to classification as an association.[41] Because failure to meet the section 761 prohibition against joint sales indicates a joint profit motive, which is a basic corporate/partnership characteristic, care must be taken not to meet more than two of the *Kintner* regulation corporate characteristics in drafting the operating agreement.[42]

If an organization is subject to the partnership provisions of the Code, it appears at first glance that the same special allocation problems encountered with limited partnerships will attend. The IRS position is that if the carrying

---

36. TEX. REV. CIV. STAT. ANN. art. 6132b, § 6(1) (Vernon 1970).

37. Treas. Reg. § 1.761-1(a), T.D. 7208, 37 Fed. Reg. 20,686 (1972).

38. Bentex Oil Corp. v. Commissioner, 20 T.C. 565 (1953) (appeal dismissed).

39. 26 U.S.C.A. (I.R.C.) § 761(a) (West 1982).

40. Treas. Reg. § 1.761-2(a)(1), (3), T.D. 7208, 37 Fed. Reg. 23,161 (1972); *see supra* § 3.13(d).

41. Chapoton, *supra* note 35, at 11.

42. *See supra* § 3.14(a).

party does not own a full working interest through payout, only a percentage of the IDCs may be deducted.[43] Revenue Ruling 68-139,[44] however, which was dismissed as unreliable and "devoid of precedential value,"[45] has been resurrected by a careful rereading. To review the facts of that ruling, *B* and *C* contribute \$300*x* each for drilling and are given the total deduction for the IDCs, and *A* contributes the lease. *B* and *C* each have a 3/32 working interest and *A* a 26/32 working interest from the outset. This special allocation of IDCs was approved. The holding in this ruling describes the terms of a partnership agreement, but further analysis reveals that the organization described is in fact a "tax partnership" that has not elected to be excluded from the partnership tax provisions of the Code pursuant to section 761(a). Thus the ruling appears to approve the old carried-interest form of allocation regardless of ownership of the working interest. In the last few years, a "tax partnership" has been used frequently to circumvent the disallowance of IDCs in the non-payout carried-interest situation.[46] Other advantages of the tax partnership are that a detailed partnership agreement does not have to be drawn, co-owners may take their production in kind, and limited partners may avoid possible continuing management fees charged to the limited partnership.[47]

Potential problems seem inevitable with the tax partnership. Although the many cases and rulings that have created problems for special allocations are directed toward partnerships, the 1981 rulings discussed above in fact concerned tax partnerships subject to all the scrutiny of partnership special allocation rules.[48] Practitioners have now begun to provide for maintenance of capital accounts in joint operating agreements, as well as in partnerships.[49]

(c) The Joint Venture

Another solution to the special allocation problem is found in a recent proposal.[50] This drilling fund was proposed as a joint venture that elects exclusion from the partnership tax provisions pursuant to section 761(a).[51] The resulting organization will resemble the historic carried-interest arrangement but lack the advantage of the interpretation provided by Revenue Ruling 68-139 as discussed at *supra* section 4.04(b). In this proposal, however,

43. Treas. Reg. § 1.612-4(a), T.D. 6836, 30 Fed. Reg. 8902 (1965).

44. 1968-1 C.B. 311.

45. DeBerry, *supra* note 11 (§ 4.03(b) herein), at 408.

46. *Id.* at 422.

47. Klein, *Oil and Gas Ventures (pt. 2)*, 26 Oil & Gas Tax Q. 314, 315 (1977).

48. Letter Rul. 8133021, *supra* note 28; Letter Rul. 8133028, *supra* note 26.

49. DeBerry, *supra* note 11 (§ 4.03(b) herein), at 426, 427.

50. Shepherd & Hicks, 1980 Oil and Gas Program, Descriptive Memorandum.

51. 26 U.S.C.A. (I.R.C.) § 761(a) (West 1982).

while the sponsor assigns 100 percent of the working interest to the venturers, he participates in first revenue by retaining a 12.5 percent net overriding royalty interest, which is converted to a 25 percent working interest upon payout. Because the entire working interest (an override is *not* a working interest) is owned by the party claiming the IDCs for the complete payout period, prior IRS rulings are satisfied. Further, one of the venturers is a limited partnership, so that an investor can take advantage of the ease of reporting and the limited liability provided by this form of organization. This proposal appears to present an adequate solution to the special allocation problem but, depending on the size of the override, may be less desirable to investors. One potential problem is that the change in sharing of income at payout could result in a taxable event.[52] This result seems more likely in view of the holding in Revenue Ruling 77-176,[53] in which *L* assigns *D* (1) 100 percent working interest in a part of the lease, subject to an override that can be converted into a 50 percent working interest upon payout, and (2) a 50 percent working interest in the rest of the acreage. The IRS approved the deduction of IDCs by *D* but held that *D* must recognize taxable income for the value of the 50 percent interest in the outside acreage. While the holding does not disapprove the joint venture format, because the conversion of the override was not questioned, the ruling is nevertheless ominous because the drilling fund considered in this ruling was also that of coowners who had elected to be excluded from the application of subchapter K.

§ 4.05 Conclusion

Because the maximum deduction of IDCs is crucial to an investor in a drilling fund, the problem of special allocations continues to concern the sponsors of these funds. Remarks by former IRS Commissioner Jerome Kurtz indicate that the IRS is indeed stepping up its audit efforts in an attempt to eliminate the use of what it terms "abusive" tax shelters, and thus special allocations can expect a critical focus.[54] In an American Broadcasting Company special on tax shelters on April 15, 1981, the new IRS Commissioner, Roscoe Egger, said that the focus on tax shelters will continue and that the formula for determining which returns are audited will be studied so that potentially abusive shelters will be selected for audit. Further evidence of Egger's intent to attack abusive shelters is the enactment of new penalties for

52. Klein, *supra* note 47 (§ 4.04(b) herein), at 321.

53. 1977-1 C.B. 77.

54. Kurtz, *Abusive Tax Shelters,* TAX NOTES, Feb. 18, 1980, at 213.

promotion of "abusive tax shelters."[55] The new law imposes a penalty on sponsors who misstate the tax implications of an investment or who grossly overvalue any property or service in connection with the investment.[56]

II. Depletion Deduction

§ 4.11 Significance of Depletion Deduction

The depletion deduction is the second most crucial aspect of investment in an oil and gas drilling fund. While deduction of IDCs provides an early incentive for investment and the deduction is available whether the well is good or not, depletion deductions are available once the drilling fund obtains production. The purpose of the depletion deduction is to provide the owner of a mineral property what depreciation provides to the owner of a depreciable asset.[1] Depletion deductions enable the owner to recover the costs of a mineral deposit over the period that the mineral is depleted. There are two methods of calculating depletion, cost depletion[2] and percentage depletion,[3] and the taxpayer must use the higher of the two. Therefore, both must be calculated. Percentage depletion is limited to 50 percent of the taxable income from the property, computed without allowance for depletion.[4]

§ 4.12 Cost Depletion

Cost depletion allows the deduction of capitalized costs (such as lease bonuses, other lease acquisition costs, exploratory charges, legal fees, and certain other capital costs) of a producing property over its life, by an annual deduction computed on the basis of the actual oil and gas produced each year in relation to the estimated recoverable oil and gas. Cost depletion is determined by dividing the portion of the adjusted basis of the property allocable to the depletable asset by the estimated barrels in the reservoir at the end of the tax year.[5] The resulting unit cost is then multiplied by the number of units sold during the year, and the resulting figure equals the cost depletion available that year.[6] Estimates of the reserves must be recalculated each

---

55. Tax Equity and Fiscal Responsibility Act of 1982, Pub. L. No. 97-248, § 320(a), 96 Stat. 611 (to be codified at 26 U.S.C.A. § 6700).

56. *Id.*

1. [1980] 5 Fed. Tax Rep. (CCH) ¶ 3500.

2. 26 U.S.C.A. (I.R.C.) § 612 (West 1967).

3. *Id.* § 613 (1967 & Supp. 1982).

4. *Id.* § 613(a) (Supp. 1982).

5. [1980] 5 Fed Tax Rep. (CCH) ¶ 3508.

6. *Id.*

year, and the adjusted basis for that particular year must be used for the calculation.

The amount deducted by reason of cost depletion is limited by the basis of the property; that is, a taxpayer can never create a capital account deficit by cost depletion. In a drilling fund, cost depletion is rarely the higher figure in computing depletion and thus is chiefly utilized when an oil and gas property is acquired after production is established and percentage depletion is not available, as discussed at *infra* section 4.13.

§ 4.13 Percentage Depletion

(a) General Background

The percentage depletion deduction[7] is a flat percentage of gross income and may enable a taxpayer to recover a great deal more than cost[8] because it is not limited by the basis of the property. Percentage depletion began in 1926, and the justification for allowing recovery in excess of investment was that the deduction would stimulate exploration.[9] In 1969 the percentage was reduced from 27.5 percent, which had prevailed since 1926, to 22 percent, apparently because it was felt that stimulus for exploration was unwarranted to that extent.[10]

(b) Limitations on Percentage Depletion

In 1975 there was strong insistence by Congress that what was perceived as favoritism to the oil industry should be eliminated. The Tax Reduction Act of 1975 further limited percentage depletion for oil and gas wells.[11] Though the legislative history does not reflect it, many sources consulted feel that the oil and gas industry accepted the reduction in percentage depletion in return for the continuation of favorable treatment for IDCs.

Percentage depletion for domestic oil and gas wells is now allowed *only* for the following: (1) regulated natural gas,[12] (2) natural gas sold under fixed contract,[13] (3) natural gas from geopressured brine,[14] and (4) limited amounts of production by independent producers and royalty owners.[15]

---

7. 26 U.S.C.A. (I.R.C.) § 613 (West 1967 & Supp. 1982).

8. *See* [1983] 5 FED. TAX REP. (CCH) ¶ 3500.

9. *Id.* ¶ 3554, Committee Reports on Pub. L. No. 91-172 (1969).

10. *Id.*

11. Pub. L. No. 94-12, §§ 501(a), (b)(2), 89 Stat. 26 (1975).

12. 26 U.S.C.A. (I.R.C.) § 613A(b)(1)(A) (West Supp. 1982).

13. *Id.* § 613A(b)(1)(B).

14. *Id.* § 613A(b)(2).

15. *Id.* § 613A(c).

The rate allowed for regulated natural gas and natural gas sold under fixed contract is 22 percent,[16] while the rate specified for natural gas from geopressured brine is 10 percent.[17] The complex exemption allowed to independent producers is discussed in the next subsection.

(c) Exemption for Independent Producers and Royalty Owners

Under the new exemption, percentage depletion is available only to independent producers and royalty owners.[18] The exemption is available for as much of the average daily production as does not exceed 1,000 barrels of oil or 6,000,000 cubic feet of natural gas or some combination in which 6,000 cubic feet of gas equals 1 barrel of oil.[19] A taxpayer who directly or indirectly sells oil or natural gas through retail outlets for total annual gross receipts exceeding $5 million in the tax year or who refines more than 50,000 barrels in any one day during the tax year is ineligible for percentage depletion.[20] Beginning in 1981, the percentage depletion allowed independent producers on these amounts is reduced each year (see table 4.6).[21]

Depletion for secondary and tertiary recovery continues at 22 percent until 1984 and can be used first in computation of the 1,000-barrel limit.[22]

Percentage depletion is thus available to any but the largest individual investor in oil and gas drilling funds. An ordinary investor will rarely have excess production: 1,000 barrels a day times $30 per barrel equals $30,000 a day, or $10.9 million a year!

Several other special rules apply to percentage depletion. The taxpayer and the taxpayer's spouse and minor children constitute a single unit for purposes of the deduction, and, if proven property is transferred, percentage depletion is no longer permitted.[23] The rule against transfer was enacted to "prevent a proliferation of the amount of proven oil and gas reserves that might be eligible for percentage depletion when produced."[24] A property is considered proven if its principal value has been established by exploration.[25] The transfer rules are inapplicable to certain transfers, including those made by reason of death.[26] In addition to the limit on percentage depletion deductions to 50 percent of the *taxable income from a single property*,[27]

---

16. *Id.* § 613A(b)(1).
17. *Id.* § 613A(b)(2).
18. *Id.* § 613A(c).
19. *Id.*
20. *Id.* § 613A(d)(2), (4).
21. *Id.* § 613A(c)(5).
22. *Id.* § 613A(c)(6).
23. *Id.* § 613A(c)(8), (9).
24. [1980] 5 Fed. Tax Rep. (CCH) ¶ 3564A.014, Pub. L. No. 94-455 Senate Comm. Report.
25. 26 U.S.C.A. § 613A(c)(9) (West Supp. 1982).
26. *Id.*
27. *Id.* § 613(a).

TABLE 4.6

REDUCTION IN PERCENTAGE DEPLETION ALLOWED

Year	Percentage
1980	22%
1981	20%
1982	18%
1983	16%
1984 and thereafter	15%

section 613A provides a further limitation under the independent producer's exemption. A deduction under this exemption may not exceed 65 percent of the *taxable income for the investor* for the year.[28] If percentage depletion exceeds this limit, it may be carried forward and deducted in subsequent years, subject to the same 65 percent limit.[29]

III. MINIMUM TAX ON ITEMS OF TAX PREFERENCE

§ 4.21 Introduction

Because the tax breaks provided by certain Code provisions resulted in a few people escaping all tax liability, the minimum tax for tax preferences was adopted in 1969.[1] Items considered tax preferences include adjusted itemized deductions, accelerated depreciation, capital gain, depletion, and productive well IDCs.[2] These items are used in determining the alternative minimum tax for noncorporate taxpayers[3] and the corporate minimum tax.[4]

28. *Id.* § 613A(d)(1).

29. *Id.*

1. Tax Reform Act of 1969, Pub. L. No. 91-172, 83 Stat. 487.

2. 26 U.S.C.A. (I.R.C.) § 57, *as amended by* Tax Equity and Fiscal Responsibility Act of 1982, Pub. L. No. 97-248, § 201, 96 Stat. 411.

3. *Id.* § 55, *as amended by* Tax Equity and Fiscal Responsibility Act of 1982, Pub. L. No. 97-248, § 201(a), 96 Stat. 411.

4. *Id.* § 56, *as amended by* Tax Equity and Fiscal Responsibility Act of 1982, Pub. L. No. 97-248, § 201(c)(1), 96 Stat. 417.

The new alternative minimum tax for individuals applies to alternative minimum taxable income in excess of $40,000 per year on a joint return ($20,000 on a separate return and $30,000 in the case of unmarried individuals) at a rate of 20 percent.[5] The income tax liability for the year will be the greater of the alternative minimum tax or the regular tax computed under other provisions of the Code.[6]

§ 4.22 Intangible Drilling Costs

The excess tax preference for IDCs is computed by determining the deduction that would have been available if costs incurred in drilling had been capitalized and amortized over 120 months beginning with production and subtracting that from the amount of IDCs actually deducted.[7] The resulting figure is an item of tax preference, as shown by the following example:

Actual deduction of IDCs in one year	$120,000
Less deduction in one year if IDCs were amortized over ten years	− 12,000
Excess IDCs	$108,000
Less net income from property	− 20,000
Amount of tax preference for IDCs	$ 88,000

However, the amount of excess IDCs is treated as a tax preference only to the extent that such amount exceeds taxpayer's net income from the property.[8]

§ 4.23 Percentage Depletion—Tax Preference

The amount of percentage depletion that is considered a tax preference is the amount by which the deduction exceeds the adjusted basis of the property.[9] As discussed at *supra* section 4.13(a), the taxpayer may continue to utilize percentage depletion (as opposed to cost depletion) after the basis of the property is reduced to zero, thus allowing recovery of more than the

---

5. *Id.* § 55, *as amended by* Tax Equity and Fiscal Responsibility Act of 1982, Pub. L. No. 97-248, § 201(a), 96 Stat. 411.

6. *Id.*

7. *Id.* §§ 57(a)(11)(B), (d)(1) (Supp. 1982).

8. *Id.* § 57(a)(11)(A); *see* [1983] 1 Fed. Tax Rep. (CCH) ¶ 578R.01.

9. 26 U.S.C.A. (I.R.C.) § 57(a)(8) (West Supp. 1982).

amount invested. To the extent that the recovery exceeds investment—that is, to the extent percentage depletion in any given year exceeds the basis of the property—the excess is an item of tax preference.

IV. Recapture

§ 4.31 Recapture of Intangible Drilling Costs

Before the Tax Reform Act of 1976, drilling funds had a specific advantage over real estate tax shelters.[1] In real estate syndications, the extent to which accelerated depreciation exceeds straight-line depreciation is subject to "recapture" upon sale of the property.[2] The result is that proceeds of the sale equal to this excess amount are taxed as ordinary income. With regard to drilling funds, a taxpayer before 1976 had the benefit of deferral of taxes by deducting IDCs from ordinary income and an opportunity to convert ordinary income into capital gain by taxing the gain at sale attributed to the IDC deduction as capital gain.

In 1976 the Code was amended to require the recapture of IDCs on the sale of an oil and gas property.[3] The portion of the gain subject to taxation as ordinary income (hence, recapture) is the amount by which the IDCs deducted exceed the amount that would have been deducted if the IDCs had been capitalized.[4]

V. Crude Oil Windfall Profit Tax Act of 1980

§ 4.41 General Background

The Crude Oil Windfall Profit Tax Act of 1980 (WPT Act)[1] was signed into law by President Carter April 2, 1980, retroactively effective March 1, 1980. The tax is a temporary excise tax and phases out over a thirty-three-month period beginning the later of December 31, 1987, or when the cumulative revenues raised reach $227.3 billion.[2] The WPT Act applies only to oil and

1. Close, *supra* note 31 (§ 3.14(a) herein), at 495.
2. 26 U.S.C.A. (I.R.C.) § 1245 (West 1982).
3. *Id.* § 1254.
4. *Id.* § 1254(a)(1), (4).

1. Pub. L. No. 96-223, 94 Stat. 229.
2. 26 U.S.C.A. (I.R.C.) § 4990 (West 1980). The phaseout is to begin in any event no later than January 1991.

defines three tiers for application of the tax: tier one, which is old oil; tier two, which is stripper-well oil; and tier three, which is newly discovered oil, certain heavy oil, and oil produced by tertiary production.[3] Tier three is the applicable level for oil discovered by drilling funds created after passage of the WPT Act.

In part because of the substantial negative reaction of independent producers and royalty owners to the tax and in part to stimulate exploration by independent producers, the Economic Recovery Tax Act of 1981 (1981 Act)[4] amended the WPT Act and provided a number of new incentives for independent producers and royalty owners.

§ 4.42 Applicable Tax Rates

The windfall profit tax on a barrel of taxable crude oil is the "applicable percentage" of the windfall profit on that barrel (see table 4.7).[5]

The definition of an independent producer used in determining the rate of the windfall profit tax is essentially the same as that used for percentage depletion; that is, the amount qualifying for reduced rates is 1,000 barrels and the reduced rate is not available for most retailers and refiners.[6] Furthermore, the reduced rates are not available for royalty interest owners for tiers one and two.[7] However, since drilling fund interests are almost entirely working interests that produce tier three new oil, this exclusion does not affect the investment return of most funds. A corollary is that owners of an overriding royalty interest are eligible for the reduced rate only if the interest existed on January 1, 1980, and if a binding contract existed on February 20, 1980, under which they were to convert their override to a working interest.[8]

Tier two oil includes oil that is from a stripper-well property, defined as a property from which the average per-well production has been ten barrels or less for any consecutive twelve-month period after 1972.[9] After 1982 the 1981 Act exempts from the 30 percent rate applicable to independent producers in tier two all stripper-well oil that is attributable to an independent producer's working interest, but only if the property has not been owned by anyone other than an independent producer since July 22, 1981.[10]

---

3. *Id.* § 4991.

4. Pub. L. No. 97-34, 95 Stat. 172.

5. 26 U.S.C.A. (I.R.C.) § 4987(b) (West Supp. 1982).

6. *Id.* § 4992(b), (c) (1980 & Supp. 1982).

7. *Id.* § 4992(d)(1)(D) (1980).

8. *Id.* § 4992(d)(2)(B).

9. *Id.* § 4991(d)(1)(A); 26 C.F.R. § 212.54(c) (1979); *see also supra* § 2.42.

10. 26 U.S.C.A. (I.R.C.) § 4994(g)(1), (2) (West Supp. 1982).

TABLE 4.7

CRUDE OIL WINDFALL PROFIT TAX RATES

	Large Producers and Royalty Owners	Independent Producers
Tier one—old oil	70%	50%
Tier two—stripper oil	60%	30%
Tier three—new oil		
1982	27.5%	27.5%
1983	25%	25%
1984	22.5%	22.5%
1985	20%	20%
1986 and thereafter	15%	15%
Tier three—not newly discovered oil	30%	30%

§ 4.43 Application of Rate

The windfall profit tax rate is applied after several deductions. From the removal price (defined, with certain exceptions, as the sale price)[11] of a barrel of oil, the base price, adjusted for inflation,[12] is subtracted, as well as the state severance tax attributable to the amount exceeding the base price.[13] The base price for tier three oil is generally "determined pursuant to the method prescribed by the Secretary by regulations."[14] See the following example of computing the windfall profit tax for tier three oil. Assume a base price of $16.55 and a state severance tax of 10 percent.

11. *Id.* § 4988(c) (1980).

12. Tier three oil receives an additional 2% annual kicker adjustment to provide additional incentive for exploration. *Id.* § 4989(b)(2).

13. *Id.* § 4988(a).

14. *Id.* § 4989(d)(1). That method will appear in Treas. Reg. § 51.4989-1(c). See method in Temporary Treas. Reg. § 150.4989-1(c).

Removal price	\$30.00
Adjusted base price	−18.04 (\$16.55 + 9% inflation)
Severance tax adjustment	− 1.20 (10% × [\$30.00 − \$18.04])
Crude oil windfall profit tax	\$10.76 × 30% (tier three rate) = \$3.23

§ 4.44 Net Income Limitation on Tax

The crude oil windfall profit tax on a barrel of oil cannot exceed 90 percent of the net income attributable to that barrel.[15] This limitation was inserted in order that secondary and tertiary production methods would not be unduly discouraged. Net income per barrel is determined by dividing taxable income from a property by the number of taxable barrels produced.[16] Taxable income is gross income less expenses other than percentage depletion, IDCs, and the windfall profit tax.[17] For partnerships the net income limitation calculation is made at the partnership level.[18] Computation of net income limitation is as follows:

Net income computation:		
Gross income from property (1,000 barrels at \$30)		\$30,000
Less severance taxes and operating expenses		−(27,000)
Net income from property		\$ 3,000
Windfall profit computation:		
Gross income		30,000
Less: Adjusted base price (\$18.04)	(\$18,040)	
Severance tax adjustment (10% × [\$30,000 − \$18,040])	(\$1,196)	−(19,236)
Windfall profit		\$10,764
Windfall profit tax (30% × \$10,764)		\$ 3,229
Net income limitation on windfall profit tax (90% × \$3,000)		\$ 2,700

15. 26 U.S.C.A. (I.R.C.) § 4988(b)(1) (West 1980).

16. *Id.* § 4988(b)(2).

17. *Id.* § 4988(b)(3).

18. Main, Hurdman & Cranston, *Analysis of the Crude Oil Windfall Profits Tax: With an Emphasis on the Independent Producer*, 28 Oil & Gas Tax Q. 418, 444 (1980).

§ 4.45 Other Provisions

Because of the great pressure exerted by royalty owners for whom no exemptions or reduced rates were provided in the original WPT Act, Congress included several other provisions in the 1981 Act that reduced the burden of the windfall profit tax on royalty owners. For 1981 only, a qualified royalty owner received a $2,500 windfall profit tax credit in computing federal income tax.[19] This replaced the $1,000 credit for 1980 that was added to the original bill by amendment. The definition of "qualified royalty production" does not include oil attributable to any overriding royalty interest or net profits interest that is created after June 9, 1981, out of an operating mineral interest that is proven oil or gas property on the date the interest is created, and is not created by a binding contract entered into before June 10, 1981.[20] Since 1982, the tax credit has been replaced by an exemption for qualified royalty owners from all windfall profit tax for a limited number of barrels,[21] as shown in table 4.8. One final provision of the WPT Act available for both working interest and royalty interest owners is that the tax is a deductible expense in computing federal income tax.[22]

TABLE 4.8

QUALIFIED ROYALTY OWNER'S EXEMPT PRODUCTION

Year	Barrels/Day/Quarter
1982	2
1983	2
1984	3
1985 and thereafter	3

19. 26 U.S.C.A. (I.R.C.) § 6429(c) (West Supp. 1982).
20. *Id.* § 6429(d)(2)(A), (B).
21. *Id.* § 4994(f) (1980).
22. *Id.* § 164(a)(5) (Supp. 1982).

VI. Summary

§ 4.51 Need for Adequate Tax Information

In summary, this chapter on the tax aspects of an oil and gas drilling fund covers only a few of the many tax implications. There has been no discussion of capitalization of leasehold cost, deductibility of management fees, investment tax credit, or tax treatment on the sale of an oil and gas property or the termination of a limited partnership. While deductions of IDCs and depletion deductions are the two tax aspects that are unique to an oil and gas drilling fund and crucial to the decision to invest in one, an informed investor needs to be aware of all other tax matters as well.

The securities laws require that adequate tax information must be provided to an investor, but some practitioners fear that the necessary complexity of the information renders it useless and that a one-sentence section in the prospectus, "Consult your own tax advisor prior to investment," would be more useful and realistic. Robert Mundheim, the former general counsel of the Treasury Department, addressed a securities regulations institute in January 1980 and is quoted as saying, "One of the critical elements in promoting an abusive tax shelter scheme is the tax opinion supplied by the promoter's tax attorney."[1] His theory is that the tax opinion is often no more than fraud insurance because it assures an investor that he will not be penalized for fraud if an improper deduction is taken and subsequently disallowed on audit. He describes the types of tax opinions of concern:

1. the intentionally false or totally incompetent opinion;
2. the "don't bother me with facts" opinion, which disclaims any knowledge of the accuracy of the promoter's information;
3. the nonopinion, which discusses tax-shelter law in general but never relates it to the specific transaction at issue; and
4. the "reasonable basis but you'll probably lose" opinion, which gives lengthy advice and then warns that, if challenged, the taxpayer will probably lose.[2]

These remarks further indicate the current mood of the IRS to increase surveillance of tax shelters. Mundheim warned that, while he would prefer

---

1. Mundheim, *Abusive Tax Shelters*, Tax Notes, Feb. 18, 1980, at 213.

2. *Id.* at 214.

to see the bar remedy what he considers the abusive tax opinion, "if an appropriate response is not forthcoming, we [at Treasury] may be forced to act."[3]

---

3. *Id.*

CHAPTER FIVE

Securities Law Aspects

[*Unassigned section numbers are reserved for possible use in supplementation.*]

I. Introduction

§ 5.01 Background of Securities Laws

The Securities Act of 1933[1] (the 1933 Act) and the Securities Exchange Act of 1934[2] (the 1934 Act) were passed in the wake of the depression, when public confidence in the stock market was at its lowest ebb. These "truth-in-securities" laws were enacted to protect investors from fraudulent acts and misrepresentations in the purchase and sale of securities.[3] The 1933 Act is designed to provide disclosure, chiefly through the registration process, of information deemed necessary for the investor's protection, whereas the 1934 Act is designed to establish integrity within the securities markets. A complex body of rules, regulations, and judicial precedent concerning these laws has developed over the past half century, and their application to the marketing of drilling funds is substantial. Drilling funds must comply with both federal securities laws and applicable state laws.

§ 5.02 Drilling Funds As Securities

The threshold question, of course, is whether an interest in a drilling fund is a security and hence regulated by the securities laws. The 1933 Act defines a security to include among other things:

1. a fractional undivided interest in oil, gas, or other mineral rights;
2. an investment contract; and
3. a certificate of interest or participation in any profit-sharing agreement.[4]

The transfer of an entire interest in an oil and gas lease does not ordinarily involve a security for the purpose of federal securities law unless the transaction can be termed an investment contract.[5] The Securities Act of Texas,

---

1. May 27, 1933, ch. 38, 48 Stat. 74 (title I codified at 15 U.S.C.A. §§ 77a–77aa (West 1981)).
2. June 6, 1934, ch. 404, 48 Stat. 881 (codified at 15 U.S.C.A. §§ 78a–78hh (West 1981)).
3. 1 Fed. Sec. L. Rep. (CCH) ¶ 103.
4. 15 U.S.C.A. § 77b(1) (West 1981).
5. Watson, *Federal and State Regulation of the Sale of Oil and Gas Interests*, 28 Ann. Inst. of Oil & Gas L. & Tax'n 245, 249 (1977).

however, includes within the definition of a security any interest in an oil and gas property (lease, fee, or title).[6]

An early United States Supreme Court case considered the sale of assignments of oil leases and concluded that this transaction represented the sale of securities.[7] Although the defendant corporation argued that the sale of subdivisions of a leasehold did not involve a "fractional, undivided interest," the Court indicated that such a sale, coupled with a promise to drill, was encompassed within the definition of an investment contract.[8] In *SEC v. W.J. Howey Co.*,[9] the Supreme Court articulated a test defining an investment contract, which continues to provide guidance today. The Court defined an investment contract as an investment in a common enterprise with the expectation of profits solely from the efforts of the promoter or a third party.[10] Most commentators agree that the sale of working interests, limited partnership interests, and joint venture participations, where the investor does not exercise significant control over the operation, will qualify as the sale of a security under the *Howey* definition.[11] Since a limited partner may not exercise significant control under state law or for purposes of federal income taxation, an interest in a limited partnership easily meets the definition of a security. Coownership, coupled with a joint operating agreement, presents a closer case; but, because at the bottom line a joint venture is an investment to explore for oil and gas, it seems certain that whatever the form of ownership, investments in a drilling fund will be considered a security.

§ 5.03 Registration

Once it is determined that a drilling fund comes under the purview of the 1933 Act, it becomes necessary to register the offering or comply with an exemption from registration. This requirement applies at the state level as well. Federal registration involves providing comprehensive information to the Securities and Exchange Commission and a prospectus to investors. Complex rules govern permissible timing for giving investors the prospectus and for consummating the sale of the security; in general, however, no information can be given to an investor before the registration statement is filed and

---

6. Tex. Rev. Civ. Stat. Ann. art. 581—4(A) (Vernon 1964).

7. SEC v. Joiner Corp., 320 U.S. 344, 64 S. Ct. 120, 88 L. Ed. 88 (1943).

8. *Id.* 320 U.S. at 351.

9. 328 U.S. 293, 66 S. Ct. 1100, 90 L. Ed. 1244 (1946).

10. *Id.* 328 U.S. at 300.

11. Watson, *supra* note 5, at 249; Aslanides, *Limited Partnerships—What's Next and What's Left?* 34 Bus. Law. 257, 274 (1978).

only limited information can be given to him thereafter until the statement is effective, at which time the sale may take place.[12] State registration can usually be coordinated with the requirements of federal registration.

§ 5.04 Exemptions

The 1933 Act and the various state securities acts provide several exemptions from registration under the rationale that the disclosure requirements are not as critical in some instances as in others. Major revisions to the federal exemptive scheme were made in March 1982,[13] and the states are considering a uniform system for exemptions that will be based on the federal scheme with some variations.[14] The federal revisions were made in response to the charge that the exemptive scheme was unnecessarily complicated and as a result was disproportionately burdensome for small issuers.[15]

A drilling fund that qualifies for an exemption from federal registration is not exempt from the antifraud or civil liability provisions of the federal securities laws[16] and must comply with applicable state securities laws as well.[17]

§ 5.05 Scope of Chapter

This chapter discusses the disclosure requirements for federal and state registration of a drilling fund and the advantages and disadvantages of registration.

Part III of this chapter examines the exemptions from federal registration with particular emphasis on the newly enacted Regulation D and then reviews the proposed uniform state exemptive scheme and discusses the exemptions presently available under the Texas Securities Act as well as those proposed in Texas. The last part of the chapter discusses registration requirements concerning brokers and dealers.

---

12. 15 U.S.C.A. § 77e (West 1981).

13. Sec. Act Release No. 6389, [1981–1982 Transfer Binder] FED. SEC. L. REP. (CCH) ¶ 83,106 (Mar. 8, 1982).

14. See *infra* § 5.29(a) for discussion of uniform exemptions.

15. Sec. Act Release No. 6339 [1981–1982 Transfer Binder] FED. SEC. L. REP. (CCH) ¶ 83,014 (Aug. 7, 1982).

16. 17 C.F.R. §§ 230.501–.506 (1982), Preliminary Note 1 (1 FED. SEC. L. REP. (CCH) at 2635—7 (3-17-82)).

17. *Id.* Preliminary Note 2.

II. Registration of Securities

§ 5.11 Introduction

The disclosure of information relative to the offering of oil and gas drilling funds is generally governed by the provisions of forms S-1,[1] S-2,[2] and S-3.[3] The SEC 1982 Integrated Disclosure Program adopted this three-tier registration system by revising form S-1 and adding forms S-2 and S-3.[4] Effective May 24, 1982, form S-1 is required for all issuers who have been subject to the Securities Exchange Act reporting system for less than three years.[5] The new form S-1 is essentially unchanged, but the format references the revised Regulation S-K to a much greater extent than did the old form S-1 for uniform disclosure requirements. Regulation S-X continues to govern financial disclosures.[6]

Forms S-2 and S-3 are available for issuers who have been subject to the Securities Exchange Act reporting requirements for three years or more, and these forms permit incorporation of Exchange Act reports by reference as well as delivery of streamlined information.[7] Whichever of these forms applies to the issuer of a drilling fund, an industry guide, Guide 4, provides the specific requirements for drilling fund disclosure.[8]

Form S-18 provides a simplified registration form for offers not exceeding $5 million[9] that is now available to registrants engaged in oil and gas operations.[10] Section 5.13 of this chapter examines the registration requirements of form S-18. Guidelines for disclosure in the registration of oil and gas drilling programs in Texas are provided in section 065.11 of the State Securities Board Administrative Guidelines.[11]

---

1. 17 C.F.R. § 239.11 (1982); 2 Fed. Sec. L. Rep. (CCH) ¶¶ 6021, 7121–26.
2. 17 C.F.R. § 239.12 (1982); 2 Fed. Sec. L. Rep. (CCH) ¶¶ 6022, 7141–46.
3. 17 C.F.R. § 239.13 (1982); 2 Fed. Sec. L. Rep. (CCH) ¶¶ 6023, 7151–55.
4. SEC Accounting Release No. 306, [1937–1982 Transfer Binder] Fed. Sec. L. Rep. (CCH) ¶ 72,328 (Mar. 3, 1982).
5. See discussion of reporting requirements at *infra* § 5.23(e)(2).
6. SEC Accounting Release No. 306, [1937–1982 Transfer Binder] Fed. Sec. L. Rep. (CCH) ¶ 72,328 (Mar. 3, 1982); 5 Fed. Sec. L. Rep. (CCH) ¶ 68,521.
7. SEC Accounting Release No. 306, [1937–1982 Transfer Binder] Fed. Sec. L. Rep. (CCH) ¶ 72,328 (Mar. 3, 1982).
8. 1 Fed. Sec. L. Rep. (CCH) ¶ 3828.
9. Form S-18, 2 Fed. Sec. L. Rep. (CCH) ¶¶ 6038A, 7301, 7302.
10. *See* 47 Fed. Reg. 25,126.
11. State Sec. Bd., 7 Tex. Admin. Code ch. 121 (Shepard's, Sept. 1, 1981).

§ 5.12 SEC Disclosure Requirements—Industry Guide 4[12]

(a) Summary of Program

The prospectus must begin with a summary of four of the topics that are dealt with in greater detail later in the prospectus: the terms of the offering, compensation, participation in costs and revenues, and the application of proceeds. Each summary should refer to the page number in the prospectus where the complete discussion may be found.

(b) Risk Factors

The risks should be set out in short, concise paragraphs under subcaptions, with references to other parts of the prospectus where appropriate. Risks included can concern the possible diminution of risk spreading if the fund is not fully subscribed, the possible loss of limited liability if Texas law interprets the limited partnership agreement to provide "control" by the limited partners, the limited liquidity of the investment, conflicts of interest, the volatility of oil and gas prices, and potential tax problems.

(c) Definitions

The prospectus should define all terms used in it that are in any way unique to the oil business, such as *working interest*, *exploratory well*, *farmout*, and *lease*.

(d) Terms of the Offering

The terms of the offering are included in the summary of the program, and this section of the prospectus should detail the type of entity chosen for the drilling fund, the method of obtaining subscriptions, and the offering costs.

(e) Additional Assessments

In general, an investor is never obligated to contribute additional funds upon assessment. If assessments are expected, the prospectus should state a maximum percentage of the original investment that may be requested; this percentage normally varies between 10 and 50 percent. Assessments are used to drill additional wells when a discovery is made, to rework existing wells, or

---

12. 1 Fed. Sec. L. Rep. (CCH) ¶ 3828.

to fund secondary recovery procedures. Although the investor is rarely required to make the additional investments, the prospectus should state the consequences of electing nonparticipation. The standard formula is that the parties who do participate are allowed to recover from production of the new or reworked well three times their assessment (300 percent) before the nonparticipating party may share in production revenues.

A Seventh Circuit case has caused some securities practitioners concern that additional assessments might be characterized as separate purchases and therefore require registration or an exemption from registration.[13] In this case the court was grappling with the meaning of the rule 10b-5 language that prohibits fraud "in connection with the purchase or sale of any security."[14] The general partners had failed to provide material information to the limited partners before several assessments, and the court found that each assessment constituted an investment decision and that therefore the nondisclosure was in connection with the purchase of a security.[15] The court, possibly anticipating concern by the securities bar, continued by finding, "Capital contributions used to satisfy partnership liabilities which antedated actionable deception are of course not recoverable."[16] From this language and the general tenor of the case, it does not appear that the case stands for the proposition that registration is required for an assessment.

(f) Plan of Distribution

The plan of distribution should discuss the broker-dealer arrangement and include information on the sales commission to be paid. Other offering costs should be estimated, and the plan of financing these costs described. Often a bank loan is secured for offering costs and paid from first production. The sponsor ordinarily agrees to assume liability for the loan if no production is established by the drilling fund.

(g) Proposed Activities

The section on proposed activities is the heart of the prospectus. It contains information on the type of wells to be drilled (onshore/offshore), the balance of the program (exploratory/developmental), the geographic areas of

---

13. Goodman v. Epstein, 582 F.2d 388 (7th Cir. 1978), *cert. denied*, 440 U.S. 939 (1979).

14. 17 C.F.R. § 240.10b-5 (1982); 2 FED. SEC. L. REP. (CCH) ¶ 22,725.

15. Goodman v. Epstein, 582 F.2d 388, 414 (7th Cir. 1978), *cert. denied*, 440 U.S. 939 (1979).

16. *Id.* 582 F.2d at 415.

interest (Texas [or other specific state]/United States/foreign), and the approximate number of wells to be drilled. If the drilling fund does not have an inventory of undeveloped leases and does not plan to commit any leases to the fund until after the formation of the fund, the prospectus should say so and should describe plans for acquiring leases. The North American Securities Administrators Association (NASAA), an association of state securities administrators, has formulated *Guidelines for the Registration of Oil and Gas Programs.*[17] The "Proposed Activities" section of the prospectus usually acknowledges the NASAA restrictions concerning the acquisition and transfer of leases between the sponsor, the drilling fund, and affiliated organizations.[18] The NASAA restrictions largely concern areas of potential conflicts of interest in such transactions.

If the drilling fund has already identified the prospects that are to be drilled, these prospects constitute property of the issuer. Disclosure requirements for property of the issuer fall under the provisions of item 11(b), form S-1, which require that information will be furnished in accordance with item 102 of Regulation S-K.[19] The sponsor is generally reluctant to reveal his exploration plans to the public for fear that competitors may lease other unacquired portions of the prospects and thereby drive up costs. If the sponsor fails to disclose information on prospects already acquired, however, a dissatisfied investor can claim that such failure constituted an omission of a material fact, and this may subject the sponsor to civil liability. The prospects should be described in a narrative form that is not too optimistic and that limits geologic terms to those familiar to an experienced oil and gas investor.[20] Then the prospectus should include a tabular description of the financial interests on the initial well on each prospect[21] and also on subsequent wells. Finally, maps of each prospect identifying by appropriate symbol the location of each well in the area that has been drilled to the target formation should be included. The symbols should indicate whether the wells are producing wells, abandoned wells, or dry holes.[22]

(h) Application of Proceeds

The section on application of proceeds should set forth information in tabular form regarding the amount of the proceeds that will be available for drill-

---

17. 1 Blue Sky L. Rep. (CCH) ¶¶ 5222–32.

18. *Id.* ¶ 5231, § X.C.(1)(s).

19. 12 C.F.R. § 229.102 (1982); 5 Fed. Sec. L. Rep. (CCH) ¶ 71,012.

20. Armbrust, *Private Placement of Oil & Gas Drilling Funds*, A.B.A. Inst. on Oil & Gas Sec. L. (Mar. 1980), at 25.

21. *Id.*

22. *Id.* at 26.

ing operations (i.e., initial capital less management fee, organization costs, etc.) and the proposed drilling fund expenditures (i.e., percentages for lease acquisition and for exploratory/developmental drilling). One sponsor interviewed said that he felt this information was the most critical part of the prospectus.

(i) Participation in Costs and Revenues

All allocations of costs and revenues between the sponsor and the investors should be set forth in tabular form, with appropriate footnotes when narrative information will aid in understanding. A statement of the plan for distribution of revenues should also be included.

(j) Compensation

The compensation to the sponsor in any form, either cash or property interest, must be described. Guide 4 urges a tabular presentation in addition to a narrative one, and the most desirable one reviewed breaks down the type of compensation, such as overriding royalty interest, management fee, and reimbursement of administrative overhead, on the left margin and then indicates in columns across the top if the compensation comes from the initial subscription dollars or from revenues of the drilling fund. The chart is then filled in with the appropriate percentages. An estimated breakdown of administrative overhead expenses should also be given and should include any executive salaries.

(k) Management

Disclosures relative to management are controlled by items 11(j)–11(*l*) on form S-1[23] and by items 401, 402, and 403 of Regulation S-K.[24] The sponsor should be identified and, if the sponsor is a corporation, all officers, directors, and key employees should be identified and all family relationships disclosed. Whether the sponsor is a corporation or an individual, all family relationships between the sponsor and affiliates should be disclosed.[25] A business biography, which lists the sponsor's education and qualifications and prior and current positions held by him, should be included. If the sponsor has

---

23. 2 Fed. Sec. L. Rep. (CCH) ¶ 7123.

24. 17 C.F.R. §§ 229.401–.403 (1982); 5 Fed. Sec. L. Rep. (CCH) ¶¶ 71,041, 71,042, 71,043; *see also* Sec. Act Release No. 6364, 5 Fed. Sec. L. Rep. (CCH) ¶ 70,962B (Dec. 3, 1981).

25. *See* Armbrust, *supra* note 20 (§ 5.12 (g) herein), at 27.

been involved in legal proceedings, such as bankruptcy, criminal or civil proceedings, or SEC violations within the past five years, this information must be noted. The sponsor must also indicate any beneficial stock ownerships and the percentage of the class of stock that this ownership represents. Finally, if a limited partnership form of organization is used, the net worth of a corporate sponsor should be disclosed to show compliance with the net worth requirements of Revenue Procedure 72-13.[26] Those requirements are discussed at *supra* section 3.14(d).

(l) Conflicts of Interest

All possible conflicts of interest, especially those with prior or future drilling funds organized by the same sponsor, should be pointed out. Potential contractual or lease arrangements between the sponsor and affiliates and the drilling fund, such as purchase of leases from the sponsor or drilling contracts with an affiliate, must be described.

(m) Prior Activities

Because the sponsor is considered the issuer, he must list all prior drilling fund activities in tabular form containing (1) the drilling results thereof, (2) the total investments, and (3) the total recovery of investments to date. The information must be broken down on a well-by-well, program-by-program basis.[27] This information is usually further broken down by gross wells and net wells. A net well exists when the sum of the fractional interests in gross wells equals one. Past success ratios rarely predict future success because of prospect and market variables, and the predictive value of these tables should be disclaimed.[28]

(n) Tax Aspects

A fair summary of the tax consequences of investment in the particular drilling fund, not just a general summary of the law, should be given. Areas where the law is new or unsettled, such as the windfall profit tax or special allocation problems, should be highlighted. It should be stated that a ruling has been sought on the tax classification of the organization, or, in the alternative, it should be noted if opinion of counsel is relied on.

---

26. *Id.* at 29.
27. *Id.*
28. *Id.*

(o) Competition, Market, and Regulation

It should be noted that there is a high degree of competition in the acquisition of leases having a good potential for development. Also the recent decline in oil prices and the effect of further volatility on drilling fund return should be discussed. The possibility of new controls and regulations of the price of gas might be mentioned as well.

(p) Additional Sections

A further section of the prospectus should summarize the limited partnership, joint venture, or operating agreement, as applicable, and copies of the agreement should be attached. Then lo and behold, the prospectus is done, except that the sponsor must now assure himself that the prospectus will pass muster with the state in which registration takes place.

§ 5.13 Form S–18

Originally form S-18 was available only to corporate issuers and was not available for registrants engaged in oil and gas operations. Thus, drilling funds could not use this simplified registration form. The present form[29] continues the dollar limitation of $5 million for use of form S-18[30] but expands the class of issuers to include those involved in oil and gas operations, as well as noncorporate issuers.[31]

Form S-18 calls for narrative disclosures somewhat less expansive than form S-1, although industry guides such as Guide 4 are retained.[32] Financial statements may be prepared in accordance with generally accepted accounting principles for the purpose of registration, because, in a drilling fund offering, no oil- and gas-producing activities are underway at the time of the offering. However, subsequent financial reporting will be governed by rule 4-10 of Regulation S-X.[33] Audited financial statements are required for only two

29. 2 Fed. Sec. L. Rep. (CCH) ¶¶ 7301–05.

30. 17 C.F.R. § 239.28; 2 Fed. Sec. L. Rep. (CCH) ¶ 7302.

31. Sec. Act Release No. 6388, [1981–1982 Transfer Binder] Fed. Sec. L. Rep. (CCH) ¶ 83,105 (Mar. 3, 1982); 47 Fed. Reg. 25,126.

32. 2 Fed. Sec. L. Rep. (CCH) ¶ 7302.

33. 17 C.F.R. § 210.4-10; 5 Fed. Sec. L. Rep. (CCH) ¶ 69,200. In Securities Act Release No. 6412, [Current Binder] Fed. Sec. L. Rep. (CCH) ¶ 83,231 (June 24, 1982), the SEC proposes deletion of the disclosure requirements of Regulation S-X, rule 4-10(k), and substitution of supplementary disclosure requirements as item 302 of Regulation S-K (5 Fed. Sec. L. Rep. (CCH) ¶ 71,032).

fiscal years rather than the three required in form S-1, and these requirements may be waived in certain highly unusual cases.[34]

§ 5.14 State Disclosure Requirements

Almost all states allow registration by coordination when a registration statement is filed under the 1933 Act.[35] However, most state securities laws provide not only for disclosure but also for examination of the proposed plan of business to see if it is "fair, just and equitable."[36] This examination usually follows the guidelines established by NASAA, which were mentioned at *supra* section 5.12(g), or the state administrative guidelines.[37]

The NASAA guidelines *require* a tax ruling from the IRS or opinion from qualified counsel on the tax classification of the drilling fund.[38] Suitability standards reasonably related to the risks involved are required for investors, and the guidelines list the factors considered relevant, such as educational level, prior investment experience, and net worth.[39] "Reasonable" organization and offering expenses are required, and the total of these expenses plus any management fee may not exceed 15 percent of the initial subscription.[40] Further, the guidelines state that an overriding royalty interest will be looked on with disfavor as sponsor compensation and that any combination of fees, working interest, or net profits interest as compensation must be reasonable and all relevant factors taken into account.[41] Transactions concerning acquisition and transfer of leases between the sponsor and affiliates are restricted.[42]

Finally, the NASAA guidelines state that a prospectus that is part of a registration statement declared effective by the SEC under the federal regulations will be deemed to comply with the guidelines, provided that the state administrators may require further disclosure in their discretion.[43] The Texas Securities Board Guidelines follow the NASAA guidelines substantially and are used in coordination with them.

---

34. *Id.*; Sec. Act Release No. 6388, [1981–1982 Transfer Binder] FED. SEC. L. REP. (CCH) ¶ 83,105 (Mar. 3, 1982).

35. *See, e.g.*, TEX. REV. CIV. STAT. ANN. art. 581—7(C)(1) (Vernon Supp. 1982–1983).

36. *See, e.g.*, *id.* art. 581—7(C)(2)(1964).

37. State Sec. Bd. Regulations, 7 TEX. ADMIN. CODE ch. 121 (Shepard's, Sept. 1, 1981).

38. 1 BLUE SKY REP. (CCH) ¶ 5223, § II.C.

39. *Id.* ¶ 5225.

40. *Id.* ¶ 5226, § V.A.

41. *Id.* ¶ 5226, § V.B. Sources of sponsor compensation are discussed at *supra* § 3.15(b).

42. *Id.* ¶ 5227.

43. *Id.* ¶ 5231, § X.B.

§ 5.15 Advantages and Disadvantages of Registration

The advantages of public drilling funds are that they usually raise more money and hence provide greater diversity of prospects, which spreads the risks of the program.[44] The general partner or sponsor of a public fund will usually have greater financial resources and greater experience in the oil and gas business. The chief advantage is that registered interests are more easily transferred because the securities laws place limitations on transfer of unregistered offerings, discussed in part III immediately following. Also, in a limited partnership, free transferability is often further limited by applicable state partnership law and federal tax law as discussed at *supra* section 3.14(e) on tax classification.

Several drilling funds reviewed, which are registered limited partnership funds, have, for each of the first five years, a desirable presentment right by which a limited partner may sell his interest to the general partner at the discounted value of future production.[45] Although this right is limited to purchase by the general partner of no more than 10 percent of the rights in any one year, it is nevertheless a very advantageous inclusion in the drilling fund because it enhances transferability.

One disadvantage of the public program is the high cost of registration; at present this cost can be expected to be at least $250,000, but wide variations can occur. This cost of registration may include as much as $70,000 in printing costs as well as large attorney's fees as compensation for following the drilling fund all the way through the registration process. Accounting fees can also be very large because of the audited financial statements required by registration. Another disadvantage of registration is the delay occasioned by the lengthy SEC review process and the fact that people unfamiliar with the oil and gas industry often conduct the review. The time required from starting preparation of the registration statement to the effective date may be five to six months.

III. Exemptions from Registration of Securities

§ 5.21 General Considerations

Given the high cost of registration, many drilling funds have sought an exemption from registration requirements. Even though the offering costs of an

---

44. Humphrey, *supra* note 1 (§ 3.01 herein), at 4.

45. Cola Petroleum, Ltd. 1979 Prospectus (Dec. 28, 1978), at A-20.

exempt offering are approximately $30,000–$50,000 as contrasted with $250,000 for a public fund, the risk of liability for any issuer's failure to satisfy all the conditions of an exemption is a compelling reason for a sponsor to consider registration. Generally, a drilling fund offered under an exemption is smaller than a public fund and hence has less risk-spreading capacity. In a smaller fund, however, the investor often has the advantage of closer and more personal management of the fund, because a small operator is concerned with the impact of the success of the fund on his industry reputation. Most important, though, is the fact that the investor will (or should) know the sponsor and can personally evaluate his business acumen and integrity.

The 1933 Act[1] provides a number of statutory exemptions, and these exemptions have been further defined by federal regulation. Several of the regulations were adopted to provide a "safe harbor" from the judicially interpreted and often vague statutory exemptions. For example, rule 146 and its successor, rule 506 of Regulation D, and rule 147 were adopted to provide objective criteria for the interpretation of sections 3(a)(11) and 4(2) of the 1933 Act.[2] An attempt to use the safe-harbor provisions of the regulations, however, does not act as an election and, therefore, an issuer who fails to comply with some standard of the regulations may still rely on the exemption that has been judicially interpreted.[3] The following section discusses the federal statutory scheme and its judicial interpretation where applicable, while later sections analyze the exemptive scheme provided by the regulations with particular emphasis on the newly adopted Regulation D.

§ 5.22 Federal Exemptive Scheme in General

Section 3 of the 1933 Act exempts from registration certain classes of securities, while section 4 exempts certain types of transactions. A drilling fund might utilize any one of several exemptions. Section 3(a)(11) of the Act defines one of the exempt classes as those sold only to persons who reside in a single state if the issuer does business in that state.[4] Rule 147 is the corresponding safe-harbor regulation.[5] Because reliance on rule 147 does not act as an election and an issuer failing to comply with some standard of rule 147 may rely on the judicial interpretation of the statute, both the re-

---

1. May 27, 1933, ch. 38, 48 Stat. 74 (title I codified at 15 U.S.C.A. §§ 77a–77aa (West 1981)).

2. *Id.* §§ 77c(a)(11), 77d(2).

3. 17 C.F.R. §§ 230.501–.506 (1982), Preliminary Note 3 (1 Fed. Sec. L. Rep. (CCH) at 2635—7 (3-17-82)); *id.* § 230.147 Preliminary Note 1 (2 Fed. Sec. L. Rep. (CCH) ¶ 5718C).

4. 15 U.S.C.A. § 77c(a)(11) (West 1981).

5. 17 C.F.R. § 230.147 (1982); 2 Fed. Sec. L. Rep. (CCH) ¶ 5718C.

quirements of rule 147 and the case law surrounding section 3(a)(11) are examined at *infra* section 5.26.

Section 3(b) gives the SEC general authority to exempt certain securities if the aggregate amount of the offering does not exceed $5 million.[6] Several exemptions have been established under this section, such as rules 240 and 242,[7] now superseded by Regulation D, rules 504 and 505,[8] and Regulations A[9] and B.[10]

A private-offering exemption was created by the 1933 Act in section 4(2), which exempts from registration all "transactions by an issuer not involving any public offering."[11] Rule 146[12] was the corresponding safe-harbor regulation for section 4(2) until the adoption of Regulation D, rule 506. As with rule 147, reliance on Regulation D, rule 506, does not act as an election, and an issuer may still rely on the judicial interpretation of section 4(2). The interpretation of this section of the 1933 Act created a great deal of uncertainty with issuers before the enactment of rule 146 in 1974. Rule 146 then attempted to provide objective criteria based on the standards set by the earlier case law. Because the requirements of Regulation D, rule 506, differ from the standard supplied by rule 146 and by the judicial interpretation of section 4(2), an issuer will have a more difficult time in finding a "fall-back" position under the statute if the objective criteria of rule 506 are not met. Section 5.24 of this chapter reviews in some detail the judicial interpretation of section 4(2), since an exemption based on this interpretation is still an alternative for an issuer.

The Small Business Investment Incentive Act of 1980[13] amended the 1933 Act by adding a new exemption from registration in section 4(6).[14] This exemption is available for transactions with an unlimited number of accredited investors if the aggregate offering price of the issue does not exceed $5 million.[15] The Act, in new section 2(15), defines accredited investors as certain institutional-type investors and persons who,

> on the basis of such factors as financial sophistication, net worth, knowledge, and experience in financial matters, or amount of assets under man-

---

6. 15 U.S.C.A. § 77c(b) (West 1981).

7. 17 C.F.R. §§ 230.240, .242 (rescinded effective June 30, 1982).

8. *Id.* §§ 230.504, .505 (1982); 1 FED. SEC. L. REP. (CCH) ¶¶ 2376, 2377.

9. 17 C.F.R. §§ 230.251–.264 (1982); 1 FED. SEC. L. REP. (CCH) ¶¶ 2359–2372.

10. 17 C.F.R. §§ 230.300–.346 (1982); 1 FED. SEC. L. REP. (CCH) ¶¶ 2381–2400A.

11. 15 U.S.C.A. § 77d(2) (West 1981).

12. 17 C.F.R. § 230.146 (rescinded effective June 30, 1982).

13. Oct. 21, 1980, Pub. L. No. 96-477, tit. IV, §§ 601–603, 94 Stat. 2275 (codified in scattered sections of 15 U.S.C.).

14. 15 U.S.C.A. § 77d(6) (West 1981).

15. *Id.* § 77c(b), 77d(6).

> agement qualif[y] as an accredited investor under rules and regulations which the Commission shall prescribe.[16]

Until the adoption of Regulation D in April 1982, few drilling funds used the exemption under section 4(6), because the SEC had not issued any regulations to define an "accredited investor." However, with the enactment of Regulation D, the SEC also promulgated rule 215, which provides a definition of accredited investor for section 4(6) that is the same definition used in Regulation D.[17] Table 5.1, at the end of this chapter, provides a comparison in summary form of various provisions of several of these exemptions.

§ 5.23 Regulation D

(a) Background

In 1978 the SEC began an evaluation of the impact of the securities laws on the ability of small businesses to raise capital.[18] The conclusion of this study was that both the registration requirements and the exemptive scheme of the 1933 Act were unduly complex and imposed a significant burden on small issuers.[19]

In response to the concerns voiced at the public hearings, the SEC in 1979 undertook a number of actions to reduce the burden of the securities laws on small businesses. Rule 242 was adopted in response to criticisms of rule 146, but it was limited to use by corporate and non–oil and gas issuers;[20] the aggregate amount of securities that could be offered pursuant to Regulation A was increased from $500,000 to $1,500,000,[21] and the use of a preliminary offering circular in Regulation A was permitted;[22] resale restrictions under rule 144 were relaxed,[23] and, finally, form S-18, a simplified regis-

---

16. *Id.* § 77b(15).

17. Sec. Act Release No. 6389, [1981–1982 Transfer Binder] FED. SEC. L. REP. (CCH) ¶ 83,106 (Mar. 8, 1982), *adopting* 17 C.F.R. §§ 230.215 (rule 215) *and* 230.501 (1982) (Reg. D, rule 501).

18. This evaluation began in April and May 1978, when the SEC conducted 21 days of public hearings in six different cities to determine the extent to which the burdens imposed on small businesses by the federal securities laws could be alleviated in a manner consistent with the protection of investors.

19. Sec. Act Release No. 6389, [1981–1982 Transfer Binder] FED. SEC. L. REP. (CCH) ¶ 83,106, at 84,908 (Mar. 8, 1982).

20. 17 C.F.R. § 230.242 (removed effective June 30, 1982).

21. Sec. Act Release No. 5977, [1978 Transfer Binder] FED. SEC. L. REP. (CCH) ¶ 81,710 (Sept. 11, 1978).

22. *Id.* No. 6075, [1979 Transfer Binder] ¶ 82,101 (June 1, 1979).

23. *Id.* No. 6286, [1981 Transfer Binder] ¶ 82,821 (Feb. 6, 1981).

tration for companies going public for the first time, was introduced.[24] Although originally an issuer that was a limited partnership or one whose offering was for oil and gas interests could not use form S-18, the use of this form has been expanded to include both limited partnerships and oil and gas interests.[25]

As a part of its program to relieve small issuers of the burden of the securities laws, the SEC also enacted the Small Business Investment Incentive Act of 1980.[26] This Act added the exemption in section 4(6) for offers and sales solely to accredited investors;[27] increased the aggregate amount that could be offered under section 3(b) from $2 million to $5 million;[28] and authorized the SEC to begin the process of developing a uniform exemption from registration with the several states.[29]

Despite these efforts to make the securities laws more useful in the formation of capital by small businesses, commentary to the SEC continued to criticize the complexity of the exemptive scheme.[30] As a result of the comments received and discussions with NASAA, the SEC proposed Regulation D in August 1981.[31]

(b) Overview

Regulation D is a series of six rules designated rules 501–506.[32] Rules 501 and 502 set forth general definitions and conditions applicable to each exemption provided by the regulation, and rule 503 provides a uniform notice-of-sales form for use in all exempt offerings made pursuant to Regulation D or section 4(6). Rules 504–506 enunciate three exemptions from the registration provisions of the 1933 Act, and these exemptions are effective April 15, 1982. Rules 504 and 505, which replace rules 240 and 242, respectively,

---

24. *Id.* No. 6049, [1979 Transfer Binder] ¶ 82,046 (Apr. 3, 1979).

25. *See id.* No. 6406, [Current Binder] ¶ 83,228 (June 4, 1982), *amending* 17 C.F.R. § 239.28. *See supra* § 5.13.

26. Oct. 21, 1980, Pub. L. No. 96-477, tit. IV, §§ 601–603, 94 Stat. 2275 (codified in scattered sections of 15 U.S.C.).

27. 15 U.S.C.A. § 77d(6) (West 1981). The Incentive Act also added to the 1933 Act § 2(15), which defined "accredited investor" as one of five enumerated institutional entities or any person who, on the basis of such factors as financial sophistication, net worth, knowledge, and experience in financial matters, or amount of assets under management qualifies as an accredited investor under rules and regulations that the SEC shall prescribe. *Id.* § 77b(15).

28. *Id.* § 77c(b).

29. The uniform exemptions developed under this authorization are discussed at *infra* § 5.29(a).

30. *See* Sec. Act Release No. 6389, [1981–1982 Transfer Binder] Fed. Sec. L. Rep. (CCH) ¶ 83,106, at 84,908 (Mar. 8, 1982).

31. *Id.* No. 6339, [1981–1982 Transfer Binder] ¶ 83,014 (Aug. 7, 1981).

32. 17 C.F.R. §§ 230.501–.506 (1982); 1 Fed. Sec. L. Rep. (CCH) ¶¶ 2373–78.

are adopted under the authority of section 3(b) of the Act.[33] Rule 506, which replaces rule 146, falls under the aegis of section 4(2) of the Act.[34]

(c) Preliminary Notes

There are six preliminary notes to Regulation D. The first reminds issuers that, although they might be exempt from the registration provisions of section 5 of the 1933 Act, they are nevertheless fully accountable under the antifraud and civil liability provisions of the federal securities laws.[35] This preliminary note also requires that issuers furnish any material information that is necessary to ensure that disclosure is not misleading.

Note 2[36] makes clear the need to comply with applicable state securities laws and points out several areas of anticipated differences.

Note 3[37] addresses the matter of using Regulation D as an election, as discussed at *supra* section 5.22, specifically pointing out that failure to comply with all the standards of rule 506 does not preclude reliance on the exemption provided by section 4(2). Note 4 specifies that Regulation D is available only to those who issue the securities and not to affiliates of the issuer or for resales.[38] The provisions pertaining to resale are contained in rule 144[39] and are discussed at *infra* section 5.25.

Preliminary note 5[40] indicates that Regulation D may be used for business combinations, and note 6[41] provides that technical compliance with Regulation D will not obviate registration if there is a plan or scheme to evade the registration requirements.

(d) Rule 501—Definitions

(1) Accredited Investor

The concept of accredited investor was first developed in rule 146, in which investors who purchased $150,000 of securities were not counted toward the limit of thirty-five purchasers.[42] Later in rule 242 the term *accredited*

---

33. 17 C.F.R. §§ 230.504(a), .505(a) (1982); 1 FED. SEC. L. REP. (CCH) ¶¶ 2376, 2377.

34. 17 C.F.R. § 230.506(a); 1 FED. SEC. L. REP. (CCH) ¶ 2378.

35. 17 C.F.R. §§ 230.501–.506 (1982), Preliminary Note 1 (1 FED. SEC. L. REP. (CCH), at 2635—7 (3-17-82)).

36. *Id.* Preliminary Note 2.

37. *Id.* Preliminary Note 3.

38. *Id.* Preliminary Note 4.

39. 17 C.F.R. § 230.144; 2 FED. SEC. L. REP. (CCH) ¶ 5718.

40. 17 C.F.R. §§ 230.501–.506 (1982), Preliminary Note 5 (1 FED. SEC. L. REP. (CCH), at 2635—7 (3-17-82)).

41. *Id.* Preliminary Note 6.

42. 17 C.F.R. § 230.146(g)(2)(i)(d); rule 146 was rescinded effective June 30, 1982.

persons was given a somewhat broader definition, and such persons again were not counted toward the limit of thirty-five purchasers.[43] Section 4(6), adopted in 1980, is available for certain offerings not exceeding $5 million to an unlimited number of accredited investors.[44] However, as discussed at *supra* section 5.22, a definition of an accredited investor for the purposes of section 4(6) was not adopted until Regulation D became effective.[45] The definition of an accredited investor is now the same for section 4(6) and Regulation D.

The term *accredited investor* means any person who falls within the definition or who the issuer "reasonably believes" comes within the definition at the time of the sale.[46] The first three categories in rule 501 of Regulation D include institutional and charitable investors in the definition of "accredited investor."[47] The fourth provides that any director, executive officer, or general partner of the issuer shall be considered an accredited investor.[48]

The fifth category defines as an accredited investor any person who purchases at least $150,000 of the securities, if that person's total purchase price does not exceed 20 percent of the purchaser's net worth at the time of the sale.[49] The purchase price may be paid by cash, marketable securities, an unconditional promise to pay within five years, or a cancellation of indebtedness. The installment provision is measured from the time of sale to the purchaser, which sale is defined as the moment the commitment is undertaken by the purchaser.[50] In the case of a limited partnership drilling fund, this commitment is made when the issuer accepts the subscription agreement.[51]

Several examples of the $150,000 purchaser category were given in the SEC release adopting Regulation D.[52] For instance, an investor with a net worth of $750,000 who makes a cash investment in a drilling fund of $150,000 is accredited. The same result will follow if the $150,000 is spread over five years. However a purchase of $200,000 by that same investor in cash or over five years will not qualify the investor, because the 20 percent limitation is exceeded. An investor with a net worth of $900,000 who purchases $180,000 of securities over six years and agrees to pay $150,000 in

---

43. *Id.* § 230.242(a)(1)(ii); rule 242 was rescinded effective June 30, 1982.

44. 15 U.S.C.A. § 77d(6) (West 1981).

45. 17 C.F.R. § 230.215, *adopted in* Sec. Act Release No. 6389, [1981–1982 Transfer Binder] FED. SEC. L. REP. (CCH) ¶ 83,106 (Mar. 8, 1982).

46. 17 C.F.R. § 230.501(a) (1982) (1 FED. SEC. L. REP. (CCH) ¶ 2373).

47. *Id.* § 230.501(a)(1)–(3).

48. *Id.* § 230.501(a)(4).

49. *Id.* § 230.501(a)(5); "net worth" includes "joint net worth" of purchaser and spouse.

50. Sec. Act Release No. 6389, [1981–1982 Transfer Binder] FED. SEC. L. REP. (CCH) ¶ 83,106 n.13, at 84,912 (Mar. 8, 1982).

51. *Id.*

52. *Id.* at 84,913.

the first five years is accredited because at least $150,000 is invested in five years and the $180,000 does not exceed the 20 percent net worth limitation.

The sixth category defines any *natural* person with a net worth in excess of $1 million as an accredited investor.[53] Joint net worth of the investor and spouse is accepted as a measure.[54]

Any *natural* person with income over $200,000 in each of the last two years who reasonably expects that income in the current year is defined as an accredited investor in the seventh category.[55] To avoid inclusion of federal tax law concepts, the word *income* is not defined.[56] The release adopting Regulation D notes the distinction between revenue and income and suggests this possible method for computing income: adjusted gross income increased by deductions for long-term capital gain and depletion as well as interest exclusions and partnership losses.[57]

The eighth category provides that an entity composed entirely of accredited investors is itself an accredited investor.[58]

(2) Affiliate

An affiliate of a person is one who controls, is controlled by, or is under common control with the person specified.[59] The affiliated definition is relevant to the issue of integration, which is discussed at *infra* section 5.23(e)(1).

(3) Aggregate Offering Price

The aggregate offering price is the sum of all consideration received by the issuer, including cash, services, property, notes, and cancellation of debt.[60] The issue of aggregation is relevant to the amount of securities authorized under rules 504 and 505 pursuant to the general authority of section 3(b). To determine the aggregate offering price under rules 504 and 505, the issuer must include proceeds of all securities sold within the preceding twelve months in all transactions exempt under section 3(b)[61] or in violation of the registration provisions of section 5(a). For instance, if an issuer has conducted an offering under rule 504 on June 1, 1982, the proceeds of

53. 17 C.F.R. § 230.501(a)(6) (1 FED. SEC. L. REP. (CCH) ¶ 2373).

54. *Id.*

55. *Id.* § 230.501(a)(7).

56. Sec. Act Release No. 6389, [1981–1982 Transfer Binder] FED. SEC. L. REP. (CCH) ¶ 83,106, at 84,914 (Mar. 8, 1982).

57. *Id.*

58. 17 C.F.R. § 230.501(a)(8) (1982) (1 FED. SEC. L. REP. (CCH) ¶ 2373).

59. *Id.* § 230.501(b).

60. *Id.* § 230.501(c).

61. Exemptions under § 3(b) include Regulation A, Regulation B, and rules 504 and 505 of Regulation D.

the June offering will be aggregated with a subsequent offering under rule 505 on December 1, 1982. However, if the December 1, 1982, offering is made under rule 506, it can be of unlimited amount since rule 506 is not an exemption under section 3(b) but, rather, falls under the authority of section 4(2). Had the June offering been for $5 million under rule 505, a subsequent offering for $5 million in December under section 4(6) would not be aggregated for the same reason; that is, only section 3(b) offerings are aggregated.

The SEC release adopting Regulation D recognizes the difficulty in understanding the interaction of the concepts of aggregation and integration:

> Aggregation is the principle by which an issuer determines the dollar worth of exempt sales available directly under Section 3(b) of the Securities Act. Integration is a principle under which an issuer determines overall characteristics of its offering.[62]

Principles of integration are discussed more fully at *infra* section 5.23(e)(1).

(4) Calculation of the Number of Purchasers

Both rule 505 and rule 506 have a limit of thirty-five purchasers.[63] However, under the definition of purchaser provided in Regulation D, several purchasers are excluded from the total.[64] Most importantly, all accredited investors as defined in rule 501(a) are excluded. In addition, a relative, spouse, or relative of a spouse having the same principal residence as the purchaser will be excluded. Also, a trust or estate in which a purchaser has more than a 50 percent beneficial interest is not counted toward the limit of thirty-five; nor is a corporation or other entity of which a purchaser owns more than 50 percent of the equity securities or interest.

(5) Purchaser-Representative

Rule 506 retains the "sophistication" test of its predecessor rule 146[65] and, as in rule 146, allows a purchaser to satisfy the test by using a purchaser-representative.[66] A purchaser-representative may not be an affiliate, director,

62. Sec. Act Release No. 6389, [1981–1982 Transfer Binder] Fed. Sec. L. Rep. (CCH) ¶ 83,106 n.31, at 84,918 (Mar. 8, 1982).

63. 17 C.F.R. § 230.505(b)(2)(ii) (1 Fed. Sec. L. Rep. (CCH) ¶ 2377), § 230.506(b)(2)(i) (1982) (1 Fed. Sec. L. Rep. (CCH) ¶ 2378).

64. 17 C.F.R. § 230.501(e) (1982); 1 Fed. Sec. L. Rep. (CCH) ¶ 2373.

65. For a discussion of rule 146, see *infra* § 5.23(i).

66. 17 C.F.R. § 230.506(b)(2)(ii) (1982); 1 Fed. Sec. L. Rep. (CCH) ¶ 2378. "Offeree representative" was used in rule 146. Be-

officer, or employee of the issuer except in certain defined situations, and the purchaser-representative must have the requisite knowledge and experience to assist the purchaser in evaluating the investment.[67] This representative must be acknowledged by the purchaser in writing to the issuer, and he must disclose to the purchaser in writing any material relationship between himself and the issuer.[68]

In a note following the definition of purchaser-representative, the SEC suggests that each purchaser-representative should consider the relevant broker-dealer registration requirements. These requirements are discussed at *infra* section 5.31.

(e) Rule 502—General Conditions To Be Met

(1) Integration

In order to obtain the exemptions provided by rules 504–506, all sales that are part of the same offering must satisfy the applicable conditions of the rules.[69] However, the term *offering* is not defined in the 1933 Act or in the regulations. The integration concept may be used to combine two or more offerings that the issuer intended to be considered separately and will often result in failure to satisfy the conditions of an exemptive rule. For example, integration of two offerings might result in more than thirty-five nonaccredited purchasers and thus render any exemption claimed under rule 505 or 506 void. The traditional factors considered in determining if offers and sales will be regarded as part of a larger offering were enunciated by the SEC in 1962:

1. whether the offerings are part of a single plan of financing;
2. whether the offerings involve issuance of the same class of security;
3. whether the offerings are made at or about the same time;
4. whether the same type of consideration is to be received; and
5. whether the offerings are made for the same general purpose.[70]

A safe harbor from consideration of these factors is provided for Regulation D offerings for all offers and sales that take place more than six months

---

cause the concept of offeree qualification is eliminated in rule 506, the term has been changed to "purchaser representative."

67. 17 C.F.R. § 230.501(h)(1), (2)(1982) (1 Fed. Sec. L. Rep. (CCH) ¶ 2373).

68. *Id.* § 230.501(h)(3), (4).

69. *Id.* § 230.502(a) (1 Fed. Sec. L. Rep. (CCH) ¶ 2374).

70. Sec. Act Release No. 4552, [1962 Transfer Binder] Fed. Sec. L. Rep. (CCH) ¶ 2272.

before the start of a Regulation D offering or more than six months after the termination of such offering, as long as there are no offers or sales within either of these six-month periods.[71]

Certain problems arise if the safe-harbor provisions are not available and an issuer is considering the 1962 integration factors listed above. No-action letters issued by the SEC between 1971 and 1979 in the oil and gas area are inconsistent,[72] and since 1979 the SEC has declined to issue such letters.[73] While the published letters are of small precedential value, they nevertheless are important in understanding the SEC's position on integration of oil and gas offerings.[74]

In *Helmet Petroleum Corp.*,[75] the SEC did not integrate a public partnership and a private partnership, each formed with the same general partner to explore for oil and gas. Each partnership was to explore separate prospects and to have separate bank accounts, and investment in one partnership was not to create any rights in the assets or production of the other. Further, the partnerships were not offered simultaneously.

An opposite result was reached in *JIC Drilling Companies*,[76] where two private partnerships formed with the same general partners were required to be integrated. Though offered at the same time, one partnership was to explore in West Virginia and the other in Pennsylvania. The SEC determined that the offers represented a single plan of financing of leasehold interests owned by a common source. It is possible that the SEC found a common source because preformation interests were offered, rather than interests in two separate, existing partnerships.

Two weeks after *JIC*, a no-action letter in regard to integration was issued to *Martin Exploration Co.*[77] Similarities between *JIC* and *Martin* were that the partnerships in question each had the same general partner, kept separate records, books, and bank accounts, and had separate sources of funds; also, an investment in one partnership created no right in the other. In *Martin* the drilling was all in Louisiana, although the prospects were more than twelve miles apart. The critical factors appear to be that in *Martin* the second partnership was not formed until the first had almost completed its

---

71. 17 C.F.R. § 230.502(a) (1982); 1 Fed. Sec. L. Rep. (CCH) ¶ 2374.

72. See *Integration of Partnership Offerings: A Proposal for Identifying a "Discreet Offering,"* Bus. Law., July 1982, for an excellent proposal for new integration criteria.

73. Clover Fin. Corp., SEC No-Action Letter, Mar. 5, 1979, [1979 Transfer Binder] Fed. Sec. L. Rep. (CCH) ¶ 82,091.

74. Still, *Integration of Oil and Gas Offerings*, A.B.A. Inst. on Oil & Gas Sec. L. (Nov. 1981), at 26.

75. SEC No-Action Letter, Aug. 18, 1975 (available on LEXIS).

76. SEC No-Action Letter, Aug. 23, 1976, [1976–1977 Transfer Binder] Fed. Sec. L. Rep. (CCH) ¶ 80,765.

77. *Id.* Sept. 7, 1976, ¶ 80,794.

drilling program and the prospects drilled by the second partnership were not in the sponsor's inventory at the time the first partnership was formed.

Several concepts emerge: (1) the argument that there are two distinct issuers is strengthened if the partnerships are formed before interests are offered, (2) each partnership should develop separate prospects, (3) offerings separated in time are less likely to be integrated, and (4) maintenance of entirely separate records is important.

To combine the principles of aggregation and integration, consider the following example:[78] If a rule 506 offering is made in July 1982, the integration safe harbor is not available to a subsequent rule 505 offering in December 1982, because six months have not elapsed. The proceeds from the rule 506 offering made in July will not be aggregated with the proceeds of the rule 505 offering in December, because the earlier offering was not made pursuant to section 3(b). However, the proceeds may be added together under the principles of integration using the factors outlined in the 1962 SEC release discussed above.[79] This integration, in turn, may result in the rule 505 offering exceeding the maximum dollar limitation.

(2) Information Requirements

If an issuer sells securities under rule 504 or only to accredited investors, Regulation D does not require any specific disclosure to investors.[80] A sale under rule 505 or 506 to any nonaccredited purchaser triggers certain disclosure requirements to *all* purchasers.[81] These disclosure requirements are subdivided into those for reporting and for nonreporting companies. A company is subject to the reporting requirements of the 1934 Act if it has five hundred shareholders and total assets of $3 million[82] or if it has filed a registration statement under the 1933 Act.[83] Regulation D has different disclosure requirements for reporting and nonreporting companies.

---

78. Sec. Act Release No. 6389, [1981–1982 Transfer Binder] FED. SEC. L. REP. (CCH) ¶ 83,106 n.31, at 84,918 (Mar. 8, 1982).

79. See text accompanying *supra* note 70 (this section).

80. 17 C.F.R. § 230.502(b)(1)(i) (1982) (1 FED. SEC. L. REP. (CCH) ¶ 2374).

81. *Id.* § 230.502(b)(1)(ii).

82. 15 U.S.C.A. § 78*l*(g) (West 1981); 17 C.F.R. §§ 240.12g-1, .12g-4 (1982), *as amended by* Exchange Act Release No. 18,647, [Current Binder] FED. SEC. L. REP. (CCH) ¶ 83,204 (Apr. 15, 1982). Registration under § 12(g) triggers certain reporting requirements. 15 U.S.C.A. § 78m (West 1981).

83. 15 U.S.C.A. § 78*o*(d) (West 1981). A company that is subject to § 15 (d) will generally have its obligation to file reports under that section suspended with respect to any fiscal year in which on the first day of such fiscal year it has fewer than 500 security holders and has had total assets under $3 million at the end of its three most recent fiscal years. Exch. Act Release No. 18,647, [Current Binder] FED. SEC. L. REP. (CCH) ¶ 83,204 (Apr. 15, 1982).

(A) *Nonreporting Companies.* An issuer that is a nonreporting company and is issuing up to $5 million of securities must provide the same kind of information required in part I of form S-18.[84] The drilling fund issuer or sponsor must also provide two years of financial statements, but only the most recent must be audited.[85] If a drilling fund is not formed as a limited partnership and the issuer cannot provide the financial statements without unreasonable effort or expense, then only the issuer's balance sheet, dated within 120 days of the start of the offering, must be audited.[86] A limited partnership drilling fund whose general partner-issuer cannot provide the required financial statement without unreasonable expense may furnish tax-basis financial statements.[87] The SEC has not, however, provided any guidance about what constitutes an "unreasonable effort or expense."

For offerings by nonreporting companies over $5 million, form S-18 is not available and all issuers must use part I of whichever form they are entitled to use for registration.[88] For drilling funds this is part I of form S-1, S-2, or S-3 and Industry Guide 4. Where audited financial statements are not available without unreasonable expense, the same options for financial statements given in offerings under $5 million are available.[89]

(B) *Reporting Companies.* For reporting companies, the size of the offering does not determine the disclosure requirements.[90] Instead there are several options for disclosure regardless of the size of the offering. One option is to provide each purchaser with the most recent annual report, the definitive proxy statement filed in connection with that report, and, if requested in writing, the most recent form 10-K.[91] Another option is to provide the most recent form 10-K or a form S-1 registration statement under the 1933 Act or a form 10 registration statement under the 1934 Act.[92] Under this option an issuer may prepare a separate document containing the necessary information in lieu of submitting the actual documents.

Whichever option above is chosen, the issuer must supplement the basic information with more recent reports that have been filed since those outlined in the two previous options, as well as provide a description of the securities, the use of the proceeds, and any material changes the issuer has experienced that are not disclosed in the required documents.[93] Finally, all purchasers must be provided an opportunity to ask questions of the issuer.[94]

---

84. 17 C.F.R. § 230.502(b)(2)(i)(A) (1982) (FED. SEC. L. REP. (CCH) ¶ 2374).
85. *Id.*.
86. *Id.*.
87. *Id.*.
88. *Id.* § 230.502(b)(2)(i)(B).
89. *Id.*.
90. *Id.* § 230.502(b)(2)(ii).
91. *Id.* § 230.502(b)(2)(ii)(A).
92. *Id.* § 230.502(b)(2)(ii)(B).
93. *Id.* § 230.502(b)(2)(ii)(C).
94. *Id.* § 230.502(b)(2)(v).

(3) Manner of Offering

No general solicitation or advertising is permitted in connection with Regulation D offers and sales,[95] except in certain cases under rule 504.[96] A footnote to the Regulation D proposal in August 1981 cautions that sales to a large number of accredited investors may indicate circumstances that violate the prohibition against general advertisement and solicitations.[97]

(4) Limitations on Resale

Securities purchased under Regulation D, with certain exceptions found in rule 504, are restricted securities and may not be resold unless the purchaser registers the securities or finds another exemption.[98] The issuer must exercise reasonable care to determine that each purchaser is purchasing only for his own account[99] and must inform the purchaser in writing that resale is limited.[100] A further requirement is that a legend, indicating the restricted nature of the interest, be placed on the certificate.[101] Under rule 146 there was a requirement that stop-transfer orders be given to the issuer's agent, but this requirement has been eliminated in Regulation D.

(f) Rule 503—Notice of Sales

Form D[102] must be filed fifteen days after the first sale of a security under Regulation D, with subsequent notices due each six months and thirty days after the last sale.[103] Form D, which is also used in conjunction with sales under section 4(6),[104] consists mainly of boxes to be checked.[105]

(g) Rule 504—Exemption for Limited Offers and Sales of Securities Not Exceeding $500,000

Rule 504 is under the general authority of section 3(b), which permits the SEC to exempt certain securities up to $5 million, and the rule is available to nonreporting companies for offerings not exceeding an aggregate offering

---

95. *Id.* § 230.502(c).

96. *Id.* § 230.504(b)(1)(1 Fed. Sec. L. Rep. (CCH) ¶ 2376).

97. Sec. Act Release No. 6339, [1981–1982 Transfer Binder] Fed. Sec. L. Rep. (CCH) ¶ 83,014 n.30, at 84,464 (Aug. 7, 1981).

98. 17 C.F.R. § 230.502(d) (1982) (1 Fed. Sec. L. Rep. (CCH) ¶ 2374).

99. *Id.* § 230.502(d)(1).

100. *Id.* § 230.502(d)(2).

101. *Id.* § 230.502(d)(3).

102. *Id.* § 239.500 (2 Fed. Sec. L. Rep. (CCH) ¶¶ 6049A, 7341).

103. *Id.* § 230.503(a) (1 Fed. Sec. L. Rep. (CCH) ¶ 2375).

104. *Id.* § 239.500(a)(1) (2 Fed. Sec. L. Rep. (CCH) ¶ 6049A).

105. *See* 2 Fed. Sec. L. Rep. (CCH) ¶ 7341.

price of $500,000.[106] There is no limit on the number of purchasers under rule 504, and aggregation is determined using the principles discussed at *supra* section 5.23(d)(3). Rule 504 expands the exemption provided by rule 240, which permitted sales only up to $100,000 to one hundred investors. Each term and condition of rules 501 through 503 must be met in a rule 504 offering unless all the sales are made in a state requiring registration of the offering and a disclosure document.[107] In that event, the prohibitions against general solicitation and resale will not apply. This provision was added because many state securities laws require registration of offerings up to $500,000 to an unlimited number of investors and the SEC feels that adequate investor protection is provided by a state registration system.[108] As discussed, under the general condition regarding information requirements, no specific disclosure is required under rule 504.[109]

(h) Rule 505—Exemption for Limited Offers and Sales of Securities Not Exceeding $5,000,000

Rule 505 also provides an exemption under section 3(b) of the 1933 Act,[110] and this rule replaces rule 242. Because the disclosure requirements of rule 242 referenced form S-18, which was not available for an offering of oil and gas interests until recently, the rule itself was not available for drilling funds. New rule 505 expands the class of issuers by including all reporting and nonreporting entities except for investment companies[111] and companies subject to the disqualification provisions of Regulation A.[112] The Regulation A provisions disqualify an issuer if it is subject to certain court proceedings involving violation of the securities laws; if the directors, officers, general partners, or 10 percent owners of the issuer have incurred a penalty for such violation; if an underwriter of such securities is subject to such violation; and, finally, if a reporting company issuer has not filed all the required reports with the SEC within the past twelve months.[113] The aggregate permissible offering price under rule 505 is $5 million determined by the aggregation principles discussed earlier.[114]

Unlike rule 504, there is a limit of thirty-five nonaccredited purchasers

---

106. 17 C.F.R. § 230.504(a) (1982) (1 FED. SEC. L. REP. (CCH) ¶ 2376).

107. *Id.* § 230.504(b)(1).

108. Sec. Act Release No. 6339, [1981–1982 Transfer Binder] FED. SEC. L. REP. (CCH) ¶ 83,014, at 84,467 (Aug. 7, 1981).

109. 17 C.F.R. § 230.502(b)(1)(i) (1982) (1 FED. SEC. L. REP. (CCH) ¶ 2374).

110. *Id.* § 230.505(a) (1 FED. SEC. L. REP. (CCH) ¶ 2377).

111. *Id.*.

112. *Id.* § 230.505(b)(2) (iii).

113. *Id.* § 230.252(c), (d), (e), (f) (1982) (1 FED. SEC. L. REP. (CCH) ¶ 2360).

114. *See supra* § 5.23(d)(3).

for rule 505.[115] An unlimited number of accredited purchasers is permissible. The offering must meet all the terms and conditions of rules 501 through 503; thus, the disclosure requirements are governed by Regulation D, section 502(b).[116]

(i) Rule 506—Exemption for Limited Offers and Sales Without Regard to Dollar Amount of Offering

Rule 506 provides an exemption under section 4(2) of the 1933 Act for reporting and nonreporting entities[117] and replaces rule 146. Because most exempt drilling funds used rule 146 in the past, a brief review of its provisions follows so that a clear comparison with rule 506 can be made.

Rule 146 allowed sales to no more than thirty-five purchasers and did not count certain purchasers toward that limit.[118] For example, purchasers of $150,000 or more were not included.[119] Each *offeree* had to be qualified; the standard for qualification under rule 146 was that the issuer reasonably believed that the offeree was sophisticated or able to bear the risk of the investment.[120] Before purchase, the drilling fund issuer had the burder of making reasonable inquiries to determine (1) that each offeree had the requisite experience in financial and business matters or (2) that the offeree and his offeree-representative together had such experience *and* that the offeree was able to bear the economic risk of the investment.[121]

The most significant change in rule 506 is that the qualification of offerees is no longer required. Furthermore, the qualification of purchasers is relaxed by rule 506, in that the issuer must only reasonably believe that each nonaccredited purchaser either alone or with his purchaser-representative has the knowledge and experience to evaluate the merits and risks of the drilling fund investment.[122] There is no longer a requirement to determine that the investor has the ability to bear the economic risk of the investment. The SEC made this change because it was felt that investor protection was adequately served by the required sophistication standard, disclosure, and notice of resale restrictions.[123]

The second most significant change in rule 506 is the incorporation of

---

115. 17 C.F.R. § 230.505(b)(2)(ii)(1982) (1 Fed. Sec. L. Rep. (CCH) ¶ 2377).

116. *See supra* § 5.23(e)(2).

117. 17 C.F.R. § 230.506(a) (1 Fed. Sec. L. Rep. (CCH) ¶ 2378).

118. *Id.* § 230.146(g)(1), (2) (rescinded effective June 30, 1982).

119. *Id.* § 230.146(g)(2).

120. *Id.* § 230.146(d).

121. *Id.*

122. *Id.* § 230.506(b)(2)(ii) (1 Fed. Sec. L. Rep. (CCH) ¶ 2378).

123. Sec. Act Release No. 6339, [1981–1982 Transfer Binder] Fed. Sec. L. Rep. (CCH) ¶ 83,014, at 84,469 (Aug. 7, 1981).

the accredited-investor standard.[124] While rule 146 did not count purchasers of $150,000 or more, the accredited-investor standard expands this concept considerably.[125] As in rules 504 and 505, an exemption under rule 506 must comply with all the terms and conditions of rules 501 through 503.[126]

As discussed at *supra* section 5.22, attempted reliance on rule 506 does not act as an election, and an issuer may still rely on the judicial interpretation of section 4(2). Some of the possible problems that an issuer might encounter in attempting to rely on section 4(2) after failing to meet a standard or condition of rule 506 are addressed at *infra* section 5.24(b).

(j) Conclusion

The simplification provided in Regulation D by the standardization of definitions, conditions, and forms should alleviate much of the burden of the federal exemptive scheme on small drilling fund offerings. As discussed earlier, however, an exemption under the federal law does not provide an exemption under the state securities laws. Recognizing the need for uniformity with state law if Regulation D is to provide the anticipated simplicity, the SEC has undertaken discussions with NASAA to accomplish this uniformity.[127] The interface of Regulation D with state law is reviewed at *infra* section 5.29.

Rules 504 and 505 are adopted under section 3(b) of the 1933 Act, for which there is no judicial interpretation; section 3(b) provides only a general authority to the SEC to exempt certain offerings up to $5 million. Rule 506, however, is under the aegis of section 4(2) of the 1933 Act, which exempts private offerings and has been interpreted extensively by the courts since 1953.[128] Because reliance on rule 506 does not act as an election and an issuer may still rely on the judicial interpretation of section 4(2), the next section examines this interpretation and some problems that are anticipated if an issuer is required to use a section 4(2) exemption in a situation where the terms and conditions of rule 506 are not met.

---

124. 17 C.F.R. § 230.506(b)(2)(i) (1 Fed. Sec. L. Rep. (CCH) ¶ 2378), § 230.501(e) (1982) (1 Fed. Sec. L. Rep. (CCH) ¶ 2373).

125. *See supra* § 5.23(d)(1).

126. 17 C.F.R. § 230.506(b)(1) (1982); 1 Fed. Sec. L. Rep. (CCH) ¶ 2378.

127. The coordination with NASAA is under the authority of § 19(c)(3) of the 1933 Act (15 U.S.C.A. § 77s(c)(3) (1981)).

128. SEC v. Ralston Purina Co., 346 U.S. 119, 73 S. Ct. 981, 97 L. Ed. 1494 (1953).

§ 5.24 Private–Offering Exemption Under the Securities Act

(a) Judicial Development of Statutory Exemption

The 1933 Act, section 4(2), exempts from registration all "transactions by an issuer not involving any public offering."[129] The SEC provided little guidance in interpreting this exemption in the early years, and the seminal case of *SEC v. Ralston Purina Co.*,[130] decided twenty years later, is still a touchstone for analysis of section 4(2). In that case the Court held that consideration of whether or not a security was a public offering "should turn on whether the particular class of persons affected need the protection of the Act. An offering to those who are shown to be able to fend for themselves is a transaction 'not involving any public offering.'"[131] The Court continued by stressing that access to the information that registration would disclose would be a critical factor.

In the following years the criteria established in *Ralston Purina* were interpreted to permit sales under the exemption to wealthy investors who were presumed to be sophisticated in financial matters.[132] The 1970s brought a rash of cases interpreting section 4(2), and the different circuits alternatively stressed the requirements for sophistication of the investor and for access to information. The Tenth Circuit seemed to emphasize sophistication,[133] whereas the Fourth Circuit stressed access to information.[134] The Fifth Circuit applied the strictest standards. In *Hill York Corp. v. American International Franchises, Inc.*,[135] the court held that while the requisite sophistication was present, that fact alone would not carry the day.[136] Later, in *SEC v. Continental Tobacco Co.*,[137] the court reviewed the requirements necessary to establish "access," such as the relationship of the offeree to the issuer, which might enable the investor to obtain information. This case caused considerable consternation in the securities bar,[138] because, even though Continental made extensive disclosures in an offering memorandum,

---

129. 15 U.S.C.A. § 77d(2) (1981).

130. 346 U.S. 119, 73 S. Ct. 981, 97 L. Ed. 1494 (1953).

131. *Id.* 346 U.S. at 125.

132. Arlberg & Lybecker, *New SEC Rules 146 and 147: The Non-Public and Intrastate Offering Exemption from Registration for the Sale of Securities*, 74 Colum. L. Rev. 622, 625 (1974).

133. Lively v. Hirschfeld, 440 F.2d 631 (10th Cir. 1971); Garfield v. Strain, 320 F.2d 116 (10th Cir. 1963).

134. United States v. Custer Channel Wing Corp., 376 F.2d 675 (4th Cir. 1967).

135. 448 F.2d 680 (5th Cir. 1971).

136. *Id.* at 690.

137. 463 F.2d 137 (5th Cir. 1972).

138. Arlberg & Lybecker, *supra* note 132, at 627.

the court noted that, because "the offers of common stock were to dentists, physicians, housewives and businessmen, who had no relationship with Continental other than that of shareholder,"[139] the requirement of adequate access was not met and the exemption was unavailable. This case implied that disclosure *and* sophistication were not sufficient if the investors had no prior relationship with the issuer that would permit access to information. This decision was so roundly criticized that it led to the adoption of SEC rule 146,[140] which provided a standard for private offerings. Rule 146 was rescinded June 30, 1982, and replaced by rule 506 of Regulation D.[141]

Doran v. Petroleum Management Corp.[142] involved investment by a sophisticated purchaser in an oil and gas drilling fund. Doran held a petroleum engineering degree from Texas A&M University, had a net worth in excess of $1 million, and held twenty-six other oil and gas properties worth $850,000—surely sophisticated by any standards. But, as *Hill York*[143] teaches, that is not sufficient. This fund was not the standard drilling fund but, rather, involved formation of a limited partnership after two of the four proposed wells were already producing. Doran agreed to a cash contribution plus assumption of a note owed by the general partners. The general partners deliberately overproduced the wells on the property, and the Wyoming Conservation Commission penalized them by sealing the wells for 338 days. Deprived of his share of the income from the property, Doran defaulted on the note and sued for rescission, claiming violation of section 4(2).[144] The court took the opportunity in this case to clear up the confusion created in *Continental Tobacco*[145] by holding that there was a disjunctive requirement of disclosure of information *or* access to information by virtue of a privileged relationship with the issuer.[146] The court remanded the case because, although Doran had been provided with adequate information, it had not been proved that *all* the offerees had such information, and the court held that the requirements of section 4(2) ran to offerees and purchasers alike.[147] The court implied that, while rule 146 did not control the case, the rule was nevertheless a standard against which future cases considering a private-offering exemption under section 4(2) would be measured.

---

139. 463 F.2d at 158.

140. 17 C.F.R. § 230.146 (1980) (rescinded effective June 30, 1982).

141. Sec. Act Release No. 6389, [1981–1982 Transfer Binder] FED. SEC. L. REP. (CCH) ¶ 83,106 (Mar. 8, 1982); rule 506 is found at 17 C.F.R. § 230.506 (1982).

142. 545 F.2d 893 (5th Cir. 1977).

143. Hill York Corp. v. American Int'l Franchises, Inc., 448 F.2d 680 (5th Cir. 1971).

144. 15 U.S.C.A. § 77d(2) (West 1981).

145. SEC v. Continental Tobacco Co., 463 F.2d 137 (5th Cir. 1972).

146. Doran v. Petroleum Management Corp., 545 F.2d 893, 905 (5th Cir. 1977).

147. *Id.* at 909.

(b) Interface with Rule 506

As noted in *Doran*,[148] even though section 4(2) controlled the case, the court implied that rule 146 provided a standard against which future cases considering a private-offering exemption under section 4(2) would be measured. This measure, of course, includes the qualification of each offeree. Indeed, this focus on offerees seems well in place as indicated by a case that considered an exemption under section 4(2) four years after *Doran*: "We emphasize that the offerees and not the ultimate purchasers are the proper reference group."[149]

If a drilling fund issuer attempts to gain an exemption under rule 506 but fails to have the proper number of nonaccredited investors or fails to meet some other condition or term of the rule, then his normal alternative will be to claim an exemption under section 4(2).[150] However, because the issuer will undoubtedly have failed to qualify the offerees because such qualification is not required by rule 506, it is unlikely that the section 4(2) exemption will be available. This seems an anomalous situation and one not intended by the drafters of Regulation D. Undoubtedly an early case will resolve the dilemma, but until then the failure to qualify for a section 4(2) exemption because all offerees are not qualified remains a possibility.

§ 5.25 Rule 144—Resale of Restricted Securities

Except as provided in certain instances in rule 504,[151] securities acquired in a transaction under Regulation D are restricted and cannot be resold without registration under the 1933 Act or an exemption therefrom. Rule 502 requires that a legend be placed on the certificate or other document evidencing ownership of the securities indicating that they have not been registered and the restrictions on their sale or transfer.[152] The issuer must also demonstrate that he has made reasonable inquiries to determine that the purchaser is acquiring the securities for investment purposes, and he must also provide written disclosure of the restricted nature of the securities to each purchaser.[153]

---

148. *See supra* § 5.24(a).

149. SEC v. Manus, [1981–1982 Transfer Binder] FED. SEC. L. REP. (CCH) ¶ 98,307, at 91,922 (S.D.N.Y. 1981).

150. Issuers who fail to comply with the requirements of an exemption under § 4(2) might be able to claim an exemption under § 3(a)(11) for intrastate offerings, 15 U.S.C.A. § 77c(a)(11) (West 1981).

151. 17 C.F.R. § 230.504(b)(1) (1982) (1 FED. SEC. L. REP. (CCH) ¶ 2376).

152. *Id.* § 230.502(d)(3) (1 FED. SEC. L. REP. (CCH) ¶ 2374).

153. *Id.* § 230.502(d)(1), (2).

The exemption most commonly used by an investor for resale of a restricted drilling fund interest is section 4(1) of the 1933 Act,[154] which exempts ordinary trading transactions that are entered into by those other than issuers, underwriters, or dealers. Because there are complex securities definitions of the terms *issuer*, *underwriter*, and *dealer*, rule 144[155] was promulgated in 1972 to establish the criteria for meeting the section 4(1) exemption in selling restricted securities to the public. The first requirement of adequate public information is met if the security has subsequently been registered under the 1934 Act[156] or the same information required in an Exchange Act registration[157] is available to the public.[158] If adequate information is available, the investor must also meet the holding-period requirements, and in some cases there will be a limit on the volume of securities that can be sold.[159] If the investor is not an affiliate of the sponsor or general partner, he may sell his interest if it has been paid for and held by him for two years[160] and if the sale represents no more than 1 percent of the outstanding units.[161] The sales must be in a broker's transaction, and notice must be filed with the SEC if the sale exceeds five hundred shares or has an aggregate sale price in excess of $10,000.[162] After an investor has owned a drilling fund interest for three years, if adequate information is available, it may be sold with no volume limits, broker requirements, or notice.[163] If the requirements of rule 144 cannot be met, rule 237 permits (with certain restrictions) the sale of $100,000 of restricted securities in any year if the securities have been owned for five years.[164]

The resale rules do not affect a limited partnership drilling fund too severely unless a change of circumstances requires the investor to liquidate his position. If the fund is unsuccessful, the joint venture or limited partnership is dissolved; and, if it is very successful, sale is generally undesirable. There is

---

154. 15 U.S.C.A. § 77d(1) (West 1981).

155. 17 C.F.R. § 230.144 (1982); 1 FED. SEC. L. REP. (CCH) ¶ 2705A.

156. 15 U.S.C.A. § 78*l* (West 1981). This registration is distinct from that required by the 1933 Act and is triggered when an organization reaches a certain size. *Id.* § 78*l*(g)(1).

157. 17 C.F.R. § 240.15c2-11 (1982); 3 FED. SEC. L. REP. (CCH) ¶ 25,116.

158. 17 C.F.R. § 230.144(c) (1982); 1 FED. SEC. L. REP. (CCH) ¶ 2705A. The information requirement of the 1934 Act is found in 15 U.S.C.A. § 78*l*(b), (g) (West 1981).

159. 17 C.F.R. § 230.144(d), (e) (1982) (1 FED. SEC. L. REP. (CCH) ¶ 2705A).

160. *Id.* § 230.144(d)(1).

161. *Id.* § 230.144(e)(1).

162. *Id.* § 230.144(f), (h).

163. These relaxed resale restrictions for *nonaffiliates* were adopted by an amendment to rule 144. Sec. Act Release No. 6286, [1981 Transfer Binder] FED. SEC. L. REP. (CCH) ¶ 82,821 (Feb. 6, 1981).

164. 17 C.F.R. § 230.237 (1982), *as amended in* Sec. Act Release No. 6414, [Current Binder] FED. SEC. L. REP. (CCH) ¶ 83,233.

no market for a break-even fund, and, in any event, the investor in a break-even fund has recouped expenses and cannot complain if resale is not available.

§ 5.26 The Intrastate Exemption

(a) General Considerations

Section 3 of the 1933 Act exempts from registration certain *classes of securities* (as opposed to *types of transactions* exempted in section 4), and section 3(a)(11) of the Act defines one of the exempt classes as those sold only to persons living in a single state if the issuer does business within that state and is a person resident in, or a corporation incorporated by, that state.[165] The rationale for the exemption is that, because the local investors will know enough about the local issuer and will be protected by state securities law, the protection of the federal law is not required. The development of this exemption is analogous to that of section 4(2), rule 146, and rule 506 of Regulation D[166] in that many years of interpretation by the courts and the SEC culminated in the adoption of more definitive criteria in a nonexclusive rule, in this case rule 147.[167] Whether an issuer uses the statutory exemption or rule 147, two factors must be kept in mind. Even though the intrastate exemption releases the issuer from the disclosure requirements of a registration statement, the issuer is still subject to the antifraud sections of the 1933 and 1934 Acts,[168] so that any disclosure made must be accurate and complete. Further, an exemption under section 3(a)(11) will not provide a state exemption, and so an exemption such as a state private-offering exemption will be needed. The same integration criteria discussed at *supra* section 5.23(e)(1) in connection with Regulation D offerings apply to intrastate offerings.

(b) Statutory Exemption

The basic concept of the intrastate exemption is that the entire issue of securities must be offered and sold exclusively to residents of a single state. The determination of what constitutes an "issue" is not governed by state

---

165. 15 U.S.C.A. § 77c(a)(11) (West 1981); 1 Fed. Sec. L. Rep. (CCH) ¶ 2251.

166. *See supra* § 5.22.

167. 17 C.F.R. § 230.147 (1982); 1 Fed. Sec. L. Rep. (CCH) ¶ 2253.

168. 15 U.S.C.A. §§ 77k, 77*l*(2), 77q, 78j, 78*o* (West 1981).

law.[169] Rather, the same integration factors that apply to Regulation D offerings determine what is an issue: (1) Are the offerings part of a single plan of financing? (2) Do the offerings involve issuance of the same class of security? (3) Are the offerings made at or about the same time? (4) Is the same type of consideration to be received? and (5) Are the offerings made for the same general purpose?[170]

Moreover, since the exemption is intended only for offerings that are essentially local in character, the determination of whether the issue is sold only to persons within the state is not made until the issue has actually "come to rest" in the hands of resident investors.[171] Therefore, the phrase "sold only to persons resident" as used in section 3(a)(11) does not refer merely to the initial sale by an issuer to its underwriters or even to subsequent sales to dealers.[172]

Because of the local nature of the exemption, transactions relying on section 3(a)(11) are allowed to use general advertisement if such advertisement indicates that sales will only be made to state residents.[173] Furthermore, securities thus exempt may be offered without compliance with the prospectus requirements of the 1933 Act and the attendant disclosure provisions.[174]

The residence of the issuer presents several problems, particularly if the issuer is a partnership. One SEC no-action letter indicates that the residence of the general partner is controlling because the general partner is considered the issuer.[175] The doing-business-within-the-state requirement has been interpreted alternatively as "substantial operational activities in the state"[176] and as "a predominant amount of business" transacted in the state.[177] In one instance 65 percent of the business in the state was not enough,[178] and in another 90 percent was sufficient.[179]

The domicile of the offeree will control the determination of the residence of the offeree, and this determination in turn requires proof of intent

---

169. Shaw v. United States, 131 F.2d 476, 480 (9th Cir. 1942).

170. Sec. Act Release No. 4434, 1 Fed. Sec. L. Rep. (CCH) ¶ 2272 (Dec. 6, 1961). For discussion of integration concepts, see *supra* § 5.23(e)(1).

171. *Id.* No. 1459, ¶ 2262 (May 29, 1937).

172. *Id.* No. 4434, ¶ 2272 (Dec. 6, 1961); *see* Hillsborough Investment Corp. v. SEC, 276 F.2d 665 (1st Cir. 1960).

173. Sec. Act Release No. 4434, 1 Fed. Sec. L. Rep. (CCH) ¶ 2276 (Dec. 6, 1961).

174. *Id.*

175. [1970–1971 Transfer Binder] Fed. Sec. L. Rep. (CCH) ¶ 78,044.

176. Sec. Act Release No. 4434, 1 Fed. Sec. L. Rep. (CCH) ¶ 2274 (Dec. 6, 1961).

177. Chapman v. Dunn, 414 F.2d 153, 159 (6th Cir. 1969).

178. SEC No-Action Letter, 276 Sec. Reg. & L. Rep. (BNA) C-7 (1974).

179. *Id.* 199, C-2 (1973).

to remain in the state.[180] Thus, winter visitors, students, and military personnel cannot qualify.

A leading commentator suggests that, since the one claiming the exemption bears the burden of proof in establishing the exemption, the issuer should take a number of precautions.[181] One such precaution is to ensure that all advertising and sales literature include an appropriate legend stating that the security is offered only to residents of the state. This literature should announce further that the transaction will be rescinded ab initio if the purchaser is not a resident.[182]

(c) SEC Rule 147

(1) Basic Provisions

Rule 147[183] was adopted in January 1974, in response to problems of interpretation that had arisen in connection with section 3(a)(11).[184] All conditions of rule 147 must be met to claim the exemption, and only issuers may use the rule.

In the first section of rule 147, the transactions are described as those that are "part of an issue" sold within a single state.[185] The rule provides a safe harbor for determining what is part of an issue by providing that offers and sales made outside a six-month period on either side of the rule 147 offering will not be integrated. Within the six-month period, the same integration criteria described at *supra* section 5.23(e)(1) apply.

(2) Requirements Concerning Issuer

Rule 147(c) concerns the residence and the "doing business" requirements of the issuer. Residence of an individual issuer is determined by principal residence and thus is an easier criterion to satisfy, because it does not require proof of intent to stay as in the determination of domicile required for section 3(a)(11).[186] To establish residence, a corporation or partnership that is an issuer must be organized under the laws of the state or, if not organized

---

180. Sec. Act Release No. 4434, 1 FED. SEC. L. REP. (CCH) ¶ 2274 (Dec. 6, 1961).

181. Bloomenthal, *SEC Exemption from Registration—A New Look*, 45 CIN. L. REV. 367, 384–85 (1976).

182. *Id.*

183. 17 C.F.R. § 230.147 (1982); 1 FED. SEC. L. REP. (CCH) ¶ 2253.

184. 1 FED. SEC. L. REP. (CCH) ¶ 2253.

185. 17 C.F.R. § 230.147(a) (1982) (1 FED. SEC. L. REP. (CCH) ¶ 2253).

186. *Id.* § 230.147(c)(1).

under any state law, have its principal office in the state.[187] An SEC no-action letter concludes that if the general partner of a limited partnership to be formed is not a resident of the state in which the partnership interests are to be sold by him on behalf of the proposed limited partnership, there can be no intrastate exemption.[188] The "doing business" requirement is satisfied if the issuer derives 80 percent of its gross revenue from business within the state, has 80 percent of its assets within the state, uses 80 percent of the proceeds of the rule 147 offering within the state, and has its principal office within the state.[189]

(3) Requirements Concerning Offerees

The residence of the offerees and purchasers is defined just as for the issuer except that a corporation or partnership that is an offeree need only have its principal office in the state as opposed to being organized under the laws of the state.[190] The residence requirement applies only at the time of offer and/or sale but does not permit an installment sale to qualify, because each payment is considered a separate sale and the purchaser must reside within the state for each payment.[191]

(4) Resale

The most attractive change from prior law is the "coming to rest" provision, which allows resale outside the state nine months after the date of the last sale,[192] as opposed to one year, which is the administrative rule of thumb under section 3(a)(11) itself. To effect the nine-month limitation on resales, the certificates must be legended, stop-transfer orders must be issued to the issuer's transfer agent, and a statement of residence must be obtained from each purchaser in writing.[193]

(5) Precautions

One commentator feels that the qualification of an offeree is more difficult under rule 147 than it was under rule 146.[194] A single offer to an out-of-state resident destroys the exemption under rule 147. Carney suggests several

---

187. *Id.*

188. [1976–1977 Transfer Binder] FED. SEC. L. REP. (CCH) ¶ 80,799.

189. 17 C.F.R. § 230.147(c)(2) (1982) (1 FED. SEC. L. REP. (CCH) ¶ 2253).

190. *Id.* § 230.147(d).

191. SEC No-Action Letter [1978 Transfer Binder] FED. SEC. L. REP. (CCH) ¶ 81,738.

192. 17 C.F.R. § 230.147(e) (1982) (1 FED. SEC. L. REP. (CCH) ¶ 2253).

193. *Id.* § 230.147(f).

194. Carney, *Exemptions from Securities*

precautions. The corporation or partnership can adopt a resolution authorizing only certain people to make offers and then only in writing.[195] The writing should in turn solicit only "offers" to buy so that no power of acceptance is created in the offerees. The offers to buy can then be received in writing with proof of residence, and acceptance can follow a review of the adequacy of this proof.[196]

(d) Combination with State Exemption

The exemption under rule 147 can be combined with a state private-offering exemption, which appears to be the simplest combination of exemptions if the residence and doing business conditions can be met without skewing the organization of a given drilling fund. Under this combination there is no state or federal notification requirement, no formal disclosure requirement, and no dollar limit on the drilling fund.

§ 5.27 Regulation A

Regulation A offerings fall under the general authority of section 3(b) of the 1933 Act, which allows the SEC to exempt certain securities if the aggregate amount of the offering does not exceed $5 million.[197] Although the maximum exemption allowed under section 3(b) is $5 million, the SEC has chosen to exempt only amounts up to $1.5 million under Regulation A.[198] The exemption under Regulation A is not available for the sale of fractional undivided interests in oil and gas rights[199] but is available for drilling funds invested in such interests. The exemption for sale of fractional undivided interests is provided in Regulation B, discussed at *infra* section 5.28(b). Regulation A contains four provisions that can disqualify an issuer.[200] These provisions make Regulation A unavailable for an issuer if he is subject to certain court proceedings involving violation of the securities laws; if the directors, officers, general partners, or 10 percent owners of the issuer have incurred a penalty for such violation; if an underwriter of such issuer's securities is subject to such violation; or if the issuer is a reporting company and has failed to file all the required reports with the SEC in the past twelve months.

Regulation for Small Issuers: Shifting from Full Disclosure (pt. 2), 11 LAND & WATER L. REV. 161, 202–04.

195. *Id.* at 211.

196. *Id.*

197. 15 U.S.C.A. § 77c(b) (West 1981); 1 FED. SEC. L. REP. (CCH) ¶ 2351.

198. 17 C.F.R. § 230.254 (1982); 1 FED. SEC. L. REP. (CCH) ¶ 2362.

199. 17 C.F.R. § 230.252(b)(1)(1982)(1 FED. SEC. L. REP. (CCH) ¶ 2360).

200. *Id.* § 230.252(c), (d), (e), (f).

An exemption under Regulation A is really less like an exemption than it is a mini-registration in that there are offering circular requirements and a review by the SEC before the offering is effective. The offering circular information requirements are governed by part II, form 1-A,[201] which, while somewhat less rigorous than Guide 4, nevertheless requires substantially similar disclosure. Regulation A offerings may permit uncertified financial statements[202] and are filed in the SEC regional office for review before becoming effective.[203]

A practitioner was interviewed who had a drilling fund offering under review by the SEC in Fort Worth. He saw the review as a substantial benefit of the exemption under Regulation A because the review eliminates the possibility of losing the exemption if the offering is found later not to meet all the conditions of the rules. Although Regulation A seems to provide for an automatic effective date,[204] the Fort Worth SEC Regional Office indicates that an issuer must wait for a response approving the offering. At present such response, in the form of a comment letter, takes approximately three weeks and, after changes are made, an additional three weeks are required. The regional office approves few drilling funds submitted under Regulation A without change, thus resulting in a minimum of six weeks for approval. The practitioner interviewed disclaimed the six-week projection, because his offering had been sent back and forth between Houston and Fort Worth for three months. Thus the efficiency anticipated by registration regionally rather than federally may be illusory. Other features of Regulation A are that advertising is permitted and that a preliminary offering circular may be used before the effective date.[205] Furthermore, securities obtained in a Regulation A offering are not restricted, and thus resales are possible without reliance on rule 144.

When determining if an offering meets the $1.5 million limitation, an issuer should note that all securities sold within one year before the beginning of an offering pursuant to an exemption under section 3(b) and/or securities sold in violation of section 5[206] will be aggregated with the Regulation A offering.[207] Public, private, or intrastate offerings, properly made, will not be included.

In the release proposing Regulation D,[208] the SEC noted that many revi-

---

201. 2 Fed. Sec. L. Rep. (CCH) ¶ 7327A, form 1-A.

202. *Id.* item 13.

203. 17 C.F.R. § 230.255 (1982) (1 Fed. Sec. L. Rep. (CCH) ¶ 2363).

204. *Id.*

205. *Id.* § 230.256 (1 Fed. Sec. L. Rep. (CCH) ¶ 2364).

206. 15 U.S.C.A. § 77e (West 1981).

207. 17 C.F.R. § 230.254(a)(1) (1982); 1 Fed. Sec. L. Rep. (CCH) ¶ 2362; see § 5.23(d)(3) for discussion of aggregation principles.

208. Sec. Act Release No. 6339, [1981–1982 Transfer Binder] Fed. Sec. L. Rep. (CCH) ¶ 83,014 (Aug. 7, 1981).

sions to the rules, forms, and disclosure requirements of Regulation A were made on the same date.[209] These revisions mean that an offering under Regulation A and a registration under form S-18 are now quite similar in their disclosure requirements. The SEC seems to be inclined to combine Regulation A and form S-18 into one uniform registration specifically tailored for small issuers.[210] However, the SEC realizes that such a registration may impose significant compliance burdens on a small issuer because of the reporting requirements of the 1934 Act.[211] Anticipating this problem, the SEC has amended the rules of the 1934 Act to exempt a class of smaller issuers from the registration and reporting requirements of the Act.[212]

§ 5.28 Regulation B

(a) Introduction

Regulation B[213] has been the basis for an enormous amount of bad publicity for the oil and gas business.[214] While the publicity was completely deserved, it has undoubtedly spilled over to taint offerings of legitimate drilling funds.

Regulation B provides an exemption for single offerings of fractional undivided interests in oil and gas rights.[215] The offerings that resulted in so much publicity are titled "schedule D" offerings, because schedule D is used for a Regulation B offering of nonproducing overriding royalty interests or working interests.[216] Numerous requirements, limitations, and conditions apply to a Regulation B exemption. A few of the more salient ones are discussed in the following paragraphs.

(b) Exemption Requirements

First, the aggregate amount of an offering under Regulation B cannot exceed $250,000,[217] but because there are no aggregation requirements an unlimited number of offerings can be made without integration. A schedule D offering is generally a single-well project, because a single offering cannot

---

209. *Id.* No. 6340, ¶ 83,015.

210. *Id.* No. 6339, ¶ 83,014.

211. See discussion at *supra* § 5.23(e)(2).

212. Exch. Act Release No. 18,647, [Current Binder] FED. SEC. L. REP. (CCH) ¶ 83,204 (Apr. 15, 1982).

213. 17 C.F.R. §§ 230.300–.346 (1982); 1 FED. SEC. L. REP. (CCH) ¶¶ 2381–2400A.

214. *See The Oil Slickers*, Houston Post, Nov. & Dec. 1972.

215. 17 C.F.R. §§ 230.300–.346 (1982); 1 FED. SEC. L. REP. (CCH) ¶¶ 2381–2400A.

216. 17 C.F.R. § 230.326 (1982); 1 FED. SEC. L. REP. (CCH) ¶ 2393.

217. Reg. B, rule 302(a), 17 C.F.R. § 230.302(a) (1982); 1 FED. SEC. L. REP. (CCH) ¶ 2382.

include interests in noncontiguous tracts unless all the interests offered are landowner's royalty interests (never the case in drilling funds) and the purchaser of such an interest must be entitled to the same fractional portion of oil and gas produced from each tract.[218] No exemption is available if the offeror is in fact a dealer and is not registered as a dealer.[219] Few of the fraudulent operators in Texas in the 1970s were registered even though registration was required by the 1934 Act.[220] To circumvent this abuse, a Texas court held that even if an operator did not sell a schedule D offering, just dealing with securities that should have been licensed and regulated made one a dealer subject to state licensing requirements.[221]

Regulation B, rule 310,[222] provides that at least ten days before an offering, four copies of an offering sheet must be filed with the SEC. This offering sheet is the celebrated schedule D, a short question-and-answer form that must contain a statement in bold print that the SEC has not approved or disapproved the offering.[223] A staff member of the Texas Securities Commission who was interviewed said that he felt this disclaimer had the effect of making the schedule D look official and hence promoted investor confidence. Schedule D requires no information concerning the offeror's background or financial condition. Item 34, division II of schedule D, reads as follows: "State what assurance the purchaser of any of the interests hereby offered has that the proposed well or wells will ever be drilled and completed."[224] One commentator notes that the standard response has been "the integrity and financial ability of the offeror" and that this statement hardly provides an adequate basis for the investor to evaluate the sponsor's financial strength.[225] One practitioner interviewed said that schedule D is tantamount to a written license to steal.

In 1975 Regulation B, rule 310,[226] was amended under pressure from the states. Rule 310 now requires that state securities administrators must be notified of schedule D offerings so that states may monitor these offerings more carefully.[227] The states had actually requested reforms in the offerings

---

218. Reg. B, rule 304, 17 C.F.R. § 230.304 (1982); 1 FED. SEC. L. REP. (CCH) ¶ 2383.

219. Reg. B, rule 306(b), 17 C.F.R. § 230.306(b) (1982); 1 FED. SEC. L. REP. (CCH) ¶ 2384.

220. 15 U.S.C.A. § 78*o* (West 1981).

221. Enntex Oil & Gas Co. v. State, 560 S.W.2d 494 (Tex. Civ. App.—Texarkana 1977, writ ref'd n.r.e.).

222. 17 C.F.R. § 230.310 (1982); 1 FED. SEC. L. REP. (CCH) ¶ 2385.

223. 1 FED. SEC. L. REP. (CCH) ¶ 2419, sched. D, division I, No. 2.

224. 1 FED. SEC. L. REP. (CCH) ¶ 2420.

225. Watson, *supra* note 5 (§ 5.02 herein), at 258.

226. 17 C.F.R. § 230.310 (1982); 1 FED. SEC. L. REP. (CCH) ¶ 2385.

227. Sec. Act Release No. 5662, [1975–1976 Transfer Binder] FED. SEC. L. REP. (CCH) ¶ 80,347 (Dec. 30, 1975).

sheet as well, but the SEC was restrained when offered this opportunity to provide more regulation and, noting the allegations of abuse of schedule D, merely took the opportunity "to caution offerors [under Regulation B] regarding their responsibilities under the federal securities laws."[228] A mild rebuke indeed!

Regulation B does not provide the offeror with an exemption under state law. Because the average size of investments in a schedule D offering is only $1,000– $2,000, presumably the number of investors would usually exceed that permitted for a private offering.[229] Offerors circumvented Texas securities laws, however, by making all sales to out-of-state residents. Thus, technically no "sales" were made in Texas to bring the transactions within the purview of state law. This ruse was counteracted in *Rio Grande Oil Co. v. State*,[230] when the court held that the terms *sales*, *offer for sale*, and *sell* used in section 4(E) of the Texas Securities Act[231] should be interpreted broadly and "may be any link in the chain of the selling process."[232] The court noted that Texas was damaged if its own citizens were permitted to engage in fraudulent practices even though the persons injured were beyond the state.[233]

(c) Abuses of Regulation B

The modus operandi of the schedule D operators can be discerned from a reading of the newspaper file maintained by the Texas Securities Commission. There was generally a boiler-room approach in these offerings, with operators recruiting salesmen who possessed no background in the oil and gas business and giving them a canned sales pitch including a number of high-pressure techniques, such as promising a fabulous return on investment and indicating that the salesman just *might* be able to find the investor a place in the investment. A Dun and Bradstreet list was often used for calls. If a prospect indicated any interest, a slick brochure was sent; the offering sheet or schedule D was often not provided until the consummation of the sale. The fraudulent companies would offer interests in oil and gas areas abandoned by others as unproductive or considered marginal by legitimate companies. They would then drill a few feet, abandon the well, and report a dry hole. Because the sales contract had no arrangement for repayment of unex-

---

228. *Id.*

229. TEX. REV. CIV. STAT. ANN. arts. 581—5(I), (Q) (Vernon Supp. 1982–1983).

230. 539 S.W.2d 917 (Tex. Civ. App.—Houston [1st Dist.] 1976, writ ref'd n.r.e.).

231. TEX. REV. CIV. STAT. ANN. art. 581—4(E) (Vernon Supp. 1982–1983).

232. 539 S.W.2d at 922.

233. *Id.* at 921.

pended funds, they were pocketed by the operator. If an investor chose to visit his investment, some companies had beautiful set-ups that looked like producing wells but were in fact dummies. Out-of-state investors were obviously preferable because they seldom visited. In one instance the operator told investors he had hit oil but would need additional completion money. Once the operator secured the money, he sent a report of natural disasters (high water was typical) that had destroyed the wells! One promotional scheme touted the use of a MISER (microwave-induced stimulation of emitted reradiation), which one could use to discover oil while driving across the land—truly a Texas-size scheme!

Persons interviewed at the Texas Securities Commission say that notices of Regulation B offerings now are few and far between, although they admit that unreported schedule D offerings are still around. One author speculates that an honest operator would hesitate to use a schedule D offering because there would be a certain amount of guilt by association and the offering would surely attract investigation by state and federal securities authorities.[234]

§ 5.29 State Exemptive Scheme

(a) Development of Uniform Federal-State Limited-Offering Exemption

(1) In General

Under the authorization of section 19 of the 1933 Act,[235] the SEC has coordinated the adoption of Regulation D with NASAA.[236] The objective of this process has been to produce a uniform exemptive scheme that will apply at both the federal and state levels. In October 1981, NASAA adopted a uniform limited-offering exemption as an official policy guideline.[237] This guideline differed from Regulation D as it was ultimately adopted,[238] and NASAA then issued another uniform exemption as an official policy guideline on April 30,

234. Hubert, *Some Current Methods of Financing Oil and Gas Exploration and Development*, 25 OIL & GAS TAX Q. 320, 339 (1977).

235. 15 U.S.C.A. § 77s(c)(3) (West 1981).

236. NASAA is a voluntary organization composed of securities regulatory agencies of 49 states, the Commonwealth of Puerto Rico, and Guam, as well as Mexico and 13 provinces of Canada.

237. An official policy guideline of NASAA represents endorsement of a principle that NASAA believes has general application. NASAA has no power to enact legislation, promulgate regulations, or otherwise bind the legislatures or administrative agencies of its members.

238. Sec. Act Release No. 6389, [1981–1982 Transfer Binder] FED. SEC. L. REP. (CCH) ¶ 83,106 (Mar. 8, 1982).

1982.[239] This uniform exemption is based on rule 505 of Regulation D but encompasses features of rule 506 with certain additional terms.

(2) NASAA Uniform Exemptive Scheme

Rule 505 of Regulation D exempts offerings not exceeding $5 million if there are no more than thirty-five purchasers, and accredited investors do not count toward the limit of thirty-five.[240] Rule 506 has no dollar limit if there are no more than thirty-five nonaccredited purchasers and provides a sophistication standard for such nonaccredited purchasers; that is, the purchaser either alone or with his purchaser-representative must be able to evaluate the merits and risks of the investment.[241] In addition to the terms of rules 505 and 506, the NASAA exemption prohibits commissions in connection with the sales of securities in reliance on the exemption unless the salesperson is a registered broker-dealer.[242] Rule 505 disqualifies certain issuers for involvement in violations of the federal securities laws.[243] The NASAA exemption extends this disqualification to those involved in certain state securities law violations.[244]

In sales to nonaccredited investors, the uniform exemption adds a suitability standard[245] as well as the sophistication standard of rule 506.[246] For accredited investors qualified under the purchase amount, net worth, or income exclusion of Regulation D,[247] the uniform exemption adds only the sophistication requirement.[248]

The effect of the NASAA scheme is to combine rules 505 and 506, adding the suitability standard for nonaccredited investors and the sophistication standard to both accredited and nonaccredited investors as well as limiting the maximum exemption to offerings not exceeding $5 million. Because the NASAA guidelines represent only an endorsement of principle and are not binding on the states, it will be necessary for the various states to enact the NASAA exemption if uniformity between the federal and state ex-

---

239. 1 BLUE SKY L. REP. (CCH) ¶ 5294A.

240. 17 C.F.R. § 230.505 (1982); 1 FED. SEC. L. REP. (CCH) ¶ 2377.

241. 17 C.F.R. § 230.506 (1982); 1 FED. SEC. L. REP. (CCH) ¶ 2378.

242. NASAA Uniform Limited Offering Exemption 1.A., 1 BLUE SKY L. REP. (CCH) ¶ 5294A.

243. 17 C.F.R. § 230.505(b)(2)(iii) (1982); 1 FED. SEC. L. REP. (CCH) ¶ 2377.

244. NASAA Uniform Limited Offering Exemption 1.B., 1 BLUE SKY L. REP. (CCH) ¶ 5294A.

245. *Id.* 1.D.1.

246. 17 C.F.R. 230.506(b)(2)(ii) (1982); 1 FED. SEC. L. REP. (CCH) ¶ 2378.

247. 17 C.F.R. § 230.501(a)(5)–(7) (1982); 1 FED. SEC. L. REP. (CCH) ¶ 2373.

248. NASAA Uniform Limited Offering Exemption 1.E., 1 BLUE SKY L. REP. (CCH) ¶ 5294A.

emptive schemes is to be achieved. The following section describes the actions taken by the Texas Securities Board in response to Regulation D and the NASAA uniform limited-offering exemptions.

(b) Texas Exemptive Scheme

In late 1982 the Texas Securities Board adopted amendments to existing regulations to coordinate the federal and Texas exemptive schemes.[249] The new regulations do not adopt the NASAA official guidelines discussed at *supra* section 5.29(a) but, rather, conform Texas rules with Regulation D.

A limited-offering exemption in Texas is available if the issuer has no more than fifteen purchasers in a twelve-month period or if the total number of security holders of the issuer is not thirty-five after the sale.[250] Purchasers must be well informed and sophisticated,[251] and "accredited purchasers" are excluded from the fifteen-purchaser limitation and from the thirty-five security-holder limitation.[252] The definition of "sophisticated investor" includes the concept of a purchaser-representative as in Regulation D.[253] The definition of "accredited investor" is identical to that in Regulation D.[254]

Section 5(T) of the Texas Securities Act[255] gives the Texas Securities Board general authority to exempt transactions; under this authority the Board exempts certain offerings in which the total number of purchasers during the prior twelve-month period does not exceed thirty-five nonaccredited purchasers.[256] Accredited purchasers do not count toward this limit.[257] Purchasers under this section are not excluded from the count for purposes of section 5(I), but the section itself may be used to exceed the numbers allowed by that section.[258] The sophistication standard, which does not include the concept of purchaser-representative, applies to this exemption for both accredited and nonaccredited purchasers.[259]

The present Texas exemptive scheme relates to Regulation D rules as follows:

---

249. 7 Tex. Reg. 3777.

250. TEX. REV. CIV. STAT. ANN. art. 581—5(I) (Vernon Supp. 1982–1983). The exemption provided by § 5(Q) is for direct sales of undivided interests in oil and gas properties, and this exemption also has a limit of 35 purchasers within a 12-month period. *Id.* art. 581—5(Q).

251. State Sec. Bd., 7 Tex. Reg. 3778 (1982), *amending* 7 TEX. ADMIN. CODE § 109.4(1) (Shepard's, Sept. 1, 1982).

252. *Id.* § 109.4(1)(D).

253. *Id.* § 109.4(1)(B).

254. *Id.* 7 Tex. Reg. 3779, § 109.4(11)(G); 17 C.F.R. § 230.501(a) (1982).

255. TEX. REV. CIV. STAT. ANN. art. 581—5(T) (Vernon Supp. 1982–1983).

256. State Sec. Bd., 7 Tex. Reg. 3778 (1982), *amending* 7 TEX. ADMIN. CODE § 109.4(11) (Shepard's, Sept. 1, 1982).

257. *Id.*

258. *Id.* 7 Tex. Reg. 3779, § 109.4(11)(D).

259. *Id.* § 109.4(11)(C).

1. A rule 504 exemption for an unlimited number of purchasers is not available in Texas, because the thirty-five purchaser limit applies.
2. An exemption under Texas rule 109.4(11) combines a rule 505 and 506 exemption as follows:
   a. As in rules 505 and 506, thirty-five nonaccredited purchasers and an unlimited number of accredited purchasers are allowed.
   b. As in rule 506, a sophistication standard is applied, but the standard is applied to accredited and nonaccredited investors.
   c. As in rule 506, no dollar limit is applied.
   d. As in rule 505, certain issuers are disqualified for violation of state or federal securities laws.

IV. Broker—Dealer Registration

§ 5.31 Federal Broker–Dealer Registration

(a) Exemptions

A sponsor of an oil and gas drilling fund will often labor under the mistaken belief that exemption from the registration required by section 5 of the Securities Act of 1933 will mean automatic exemption from the broker-dealer registration requirements of the Securities Exchange Act of 1934.[1] This is not so. If a sponsor intends to market his drilling fund himself, he must register as a broker-dealer unless an exemption from such registration can be found.

The definitions of the terms *broker* and *dealer* in the 1934 Act give some clue to the SEC's interpretation of who must register. A broker is defined as "any person engaged in the business of effecting transactions in securities for the account of others,"[2] whereas a dealer is one "engaged in the business of buying and selling securities for his own account."[3] The sponsor of a drilling fund will probably not be categorized as a dealer, because he does not both buy *and* sell securities.[4] The presence of presentment rights in a drilling fund is unlikely to result in the interpretation that a sponsor is "buying" as well as "selling" securities, since presentment is both a contingent and a limited right.[5] There is no authority for this interpretation, however.

---

1. 15 U.S.C.A. § 78*o* (West 1981).
2. *Id.* § 78c(a)(4).
3. *Id.* § 78c(a)(5).
4. *See* Wertheimer & Mark, *Special Problems of Unregistered Real Estate Securities*, 22 U.C.L.A. L. Rev. 1219, 1226 (1975).
5. *See supra* § 5.15.

A sponsor can avoid the definition of broker by selling for his own account as an issuer rather than "for the account of others."[6] Thus a definitional exemption from registration as a broker-dealer has been recognized by the SEC in certain situations for the sponsor who markets his own drilling fund.[7] Two situations present themselves for consideration: in the first, the sponsor or general partner is an individual; in the second, the sponsor is a corporation that intends to use its officers, directors, and employees to sell the drilling fund. In the first situation, SEC interpretation allows the individual general partner to sell securities without classification as a broker if he does not solicit investors for recurring series of drilling funds, in which case he is no longer selling for his own account but rather is "engaged in the business of effecting transactions in securities for the account of others."[8] In the second situation, even if the sponsor of the drilling fund might avail himself of the issuer exemption, other persons selling the drilling fund interests, such as officers, directors, and employees, may be required to register if they are found to be independent contractors and not employees of the sponsor.[9] In addition to the requirement that the sales be on a nonrecurring basis, the persons selling must satisfy four other criteria: (1) they are not hired specifically to sell securities; (2) they are primarily engaged in activities other than such sales; (3) they intend to remain with the sponsor after the drilling fund is completed; and (4) they receive a fixed compensation, although receipt of commissions may not necessarily destroy the exemption.[10]

Noting that "further guidance would be useful particularly to the promoter of . . . oil and gas drilling programs," the SEC proposed a rule change in 1977 that would provide a safe harbor within which persons associated with the issuer would not be considered brokers.[11] The safe-harbor test is disjunctive and includes (1) those who confine their activities to largely intermediary activities; (2) those who are bona fide employees and satisfy certain tests, such as not having participated in selling securities for the preceding two years and being compensated by other means than commission; or

---

6. 15 U.S.C.A. § 78c(a)(4) (West 1981).

7. Exch. Act Release No. 13,195, [1976–1977 Transfer Binder] FED. SEC. L. REP. (CCH) ¶ 80,927 (Jan. 21, 1977).

8. 15 U.S.C.A. § 78c(a)(4) (West 1981); *see* SEC No-Action Letter, Sept. 26, 1975, Ginsburg, Feldman & Bress (available on LEXIS).

9. E. WEISS, REGISTRATION AND REGULATION OF BROKERS AND DEALERS § 1.2 (1965).

10. SEC No-Action Letter, Feb. 3, 1972, Woodmoor Corp., [1971–1972 Transfer Binder] FED. SEC. L. REP. (CCH) ¶ 78,653; *Examination of Staff Positions Reveals Pragmatic Approach for Sellers of Tax Shelter*, SEC. REG. & L. REP. (BNA) No. 232, C-1 (1973).

11. Exch. Act Release No. 13,195, [1976–1977 Transfer Binder] FED. SEC. L. REP. (CCH) ¶ 80,927 (Jan. 21, 1977).

(3) those whose activities are largely clerical or ministerial.[12] If adopted, this rule will have a substantial effect on sponsors of drilling funds by requiring registration by many people not currently required to do so. Although the SEC announced in 1979 that staff recommendations based on public comment would be forthcoming in 1980,[13] no response is yet available, and time augurs well for abandonment of the proposed rule.

In addition to the definitional exemption for issuers, a statutory exemption is provided for brokers and dealers whose business is exclusively intrastate.[14] This exemption has been interpreted very narrowly by the staff, whose position is that all the broker's business must be intrastate[15] and that the issuer must be considered a resident of the state and doing business there within the meaning of rule 147.[16] Importantly, qualification for the intrastate exemption does not relieve the issuer of the obligation to comply with all the rules for broker-dealers that are discussed in the next subsection.[17] For instance, the net capital and margin requirements still apply to the issuer qualifying for the intrastate exemption,[18] making attainment of this exemption a Pyrrhic victory at best.

(b) Registration Requirements

If a sponsor intends to market his own drilling fund and finds no available exemption, he must register pursuant to section 15(a)(1) of the 1934 Act[19] and comply with certain requirements. A net capital rule requires a broker-dealer to restrict his "aggregate indebtedness" to not more than 1500 percent (800 percent for the first twelve months of operation) of his "net capital" and also requires maintenance of net capital above a minimum amount.[20] Additional rules restrict a broker-dealer's use of any customer's free-credit balance and impose restrictions on a broker-dealer's use of a customer's cash left in his possession.[21] The 1934 Act also makes it unlawful for a broker-

---

12. 2 Fed. Sec. L. Rep. (CCH) ¶ 21,152.

13. SEC Release, *Updated Agenda of SEC Rule-making and Regulatory Matters*, Sec. Reg. & L. Rep. (BNA) No. 519, G-1 (Sept. 12, 1979).

14. 15 U.S.C.A. § 78*o*(a)(1) (West 1981).

15. *Examination of Staff Positions, supra* note 10, at C-2.

16. SEC No-Action Letter, Dec. 15, 1976, Allied Real Estate Sec., Inc., [1976–1977 Transfer Binder] Fed. Sec. L. Rep. (CCH) ¶ 81,041. See discussion of residence and doing business requirements under rule 147 at *supra* § 5.26(c)(2).

17. *Examination of Staff Positions, supra* note 10, at C-11.

18. *Id.*

19. 15 U.S.C.A. § 78*o* (West 1981).

20. 17 C.F.R. § 240.15c3-1(a) (1982) (3 Fed. Sec. L. Rep. (CCH) ¶ 25,126). This minimum amount is normally $25,000 but may be reduced to $5,000 in certain circumstances. *Id.* § 240.15c3-1(a)(2).

21. *Id.* §§ 240.15c3-2–.15c3-3.

dealer to extend credit to a customer[22] and requires detailed record-keeping procedures.[23]

Because the net capital and aggregate indebtedness requirements are difficult to meet and the computations exceedingly complex, many tax-shelter issuers form a new corporation to register as a broker-dealer.[24] It is recommended that this new corporation be formed as a subsidiary of the issuer and be capitalized with liquid assets and incur few liabilities in order to satisfy the net capital rule.[25] The new corporation should be limited to only a few officers and directors to ease the reporting requirements incurred with any changes.[26]

§ 5.32 State Broker–Dealer Registration

Virtually all states require licensing of persons involved in the offer or sale of securities, and all state statutes should be consulted before a drilling fund is offered in any state. Texas applies a broad judicial interpretation to the term *sell*, which is defined in the statute as "any act by which a sale is made."[27] In *Rio Grande Oil Co. v. State*,[28] the court applied the term *seller* to a person if he performed "any link in the chain of the selling process" in Texas.[29]

Texas does not make a distinction between the terms *broker* and *dealer*, using only the term *dealer* and defining it in very broad language, as follows:

> The term "dealer" shall include every person or company, other than a salesman, who engages in this state, either for all or part of his or its time, directly or through an agent, in selling, offering for sale or delivery or soliciting subscriptions to or orders for, or undertaking to dispose of, or to invite offers for, or rendering services as an investment adviser, or dealing in any other manner in any security or securities within this state.[30]

The Texas statute does, however, exempt from registration issuers finding an exemption under section 5,[31] which includes both the present and proposed private placement exemptions.[32] If a private placement is made under Reg-

---

22. 15 U.S.C.A. § 78g(c) (West 1971).

23. 17 C.F.R. § 240.17a-3 (1982); 3 Fed. Sec. L. Rep. (CCH) ¶ 26,154.

24. Wertheimer & Mark, *supra* note 4 (§ 5.31(a) herein), at 1233.

25. *Id.*

26. *Id.*

27. Tex. Rev. Civ. Stat. Ann. art. 581—4(E) (Vernon Supp. 1982–1983).

28. 539 S.W.2d 917 (Tex. Civ. App.—Houston [1st Dist.] 1976, writ ref'd n.r.e.).

29. *Id.* at 922.

30. Tex. Rev. Civ. Stat. Ann. art. 581—4(C) (Vernon 1964).

31. *Id.* art. 581—5 (Supp. 1982–1983).

32. *Id.* art. 581—4(C) (1964).

ulation D, rule 506,[33] through a registered broker-dealer, the issuer may make himself available to answer questions of investors as required by the rule and not be considered a dealer.[34]

Texas provides for a restricted registration for those dealing exclusively in the sale of oil and gas interests.[35] This particular type of registration has more limited examination and record-keeping requirements.[36]

---

33. 17 C.F.R. § 230.506 (1982); 1 FED. SEC. L. REP. (CCH) ¶ 2378.

34. State Sec. Bd., 7 TEX. ADMIN. CODE § 115.1(f) (Shepard's, Sept. 1, 1981).

35. *Id.* § 115.1(b)(1)(A).

36. *Id.* §§ 115.3(c)(3)(E), 115.5(a)(3).

TABLE 5.1

COMPARISON OF FEDERAL SECURITIES LAW EXEMPTIONS

Subject of Comparison	Rule 147	Regulation A	Section 4(6)	Rule 504	Rule 505	Rule 506
Aggregate sales price	Unlimited	$1,500,000 per offering	$5,000,000 per offering	$500,000 (12 months)	$5,000,000 (12 months)	Unlimited
Manner of offering	Solicitations must indicate sales can be made only within state of offering	Similar to registered offering	No general solicitation permitted	No general solicitation unless registered in states that require delivery of a disclosure document	No general solicitation permitted	No general solicitation permitted
Excluded issuers	Those not organized or having principal residence within state or not doing 80% of business in state or not using 80% of proceeds in state	Issuers of fractional oil and gas interests or investment companies	None	No reporting or investment companies	No investment companies or issuers disqualified under Regulation A	None
Investor qualification	Corporation or partnership must have principal office in state. Individual must have principal residence in state	No restriction	Accredited investors	None required	Accredited or none required	Purchasers must be sophisticated (alone or with representative). Accredited investors presumed to be qualified
Information requirement	No specific requirement	Offering circular with information of schedule I of form 1-A. May use preliminary offering circular if offering underwritten by registered broker-dealer	No specific requirement	No specific requirement	1. If purchased solely by accredited investors, no information specified. 2. If purchased by nonaccredited investors, a. nonreporting companies must furnish: i. offerings up to $5,000,000—information in part I of form S-18 or available registration, 2-year financials, 1 year audited—if undue effort or expense, issuers other than limited partnerships only balance sheet as of 120 days before offering must be audited—if lim-	

						expense, issuers other than limited partnerships only balance sheet as of 120 days before offering must be audited—if limited partnership and undue effort or expense, financials may be tax basis b. reporting companies must furnish: i. rule 14a-3 annual report to shareholders, definitive proxy statement and form 10-K, if requested, plus subsequent reports and other updating information *or* ii. information in most recent form S-1 or form 10 or 10-K plus subsequent reports and other updating information c. issuers must make available before sale: i. exhibits, ii. written information given to accredited investors, iii. opportunity to ask questions and receive answers.
Number of investors	No limit	No limit	No limit on number of accredited investors. No other persons	No limit	35+ unlimited accredited investors	35+ unlimited accredited investors
Resales restricted	Yes. No resale to out-of-staters until 9 months after last sale under the rule	No	Yes	Restricted unless registered in states that require delivery of a disclosure document	Yes	Yes
Report of sales	No	Advance notice of offering on form 1-A	Yes, condition of exemption	Form D required as a condition of exemption; 5 copies filed with SEC 15 days after first sale, every 6 months after first sale, 30 days after last sale		

CHAPTER SIX

Conclusion

§ 6.01 Outlook for Drilling Fund Activity

In early 1983, the United States is still a major importer of crude oil and as such is still subject to the uncertainties of world oil production policies, in particular those of OPEC. The world recession of 1982, combined with energy conservation and efficiency measures, has resulted in a significant decrease in the demand for oil and gas. OPEC has only recently made an adjustment in production to accommodate these market factors, and, chiefly as a result of OPEC's inability to reach consensus, the price of oil has dropped from $38 per barrel in 1981 to $30 in 1983. While this decline in price has helped generate economic recovery in the United States, major fluctuations in energy prices are not healthy for either the oil business or the economy in general.

The events of 1982 would seem to dictate that domestic exploration for oil and gas should continue apace in order to accomplish U.S. independence from OPEC policies. However, the number of active drilling rigs has dropped from 4,530 in December 1981 to 2,064 in March 1983.[1] A corollary to this decrease in drilling is a decrease in the number of drilling fund offerings made in 1982. Many marginal oil companies that in the past sponsored drilling funds have been driven out of the field altogether by lower oil prices that would not support a high-interest debt structure. Other potential sponsors have been unwilling to undertake drilling programs as projections for payout stretched farther and farther out on the horizon.

Investor enthusiasm for drilling funds has shown a marked decrease also as the diminished profitability curve, coupled with several changes in the tax laws, has made oil and gas tax-shelter investments less desirable. The reduction in the maximum individual tax rate from 70 percent to 50 percent has meant that tax incentives govern investor decisions to a lesser extent and potential profitability to a greater extent, and drilling fund investments now protect less income and expect less profit. Investors are also lured by the newer oil and gas income funds.[2] These funds capitalize on the financial difficulties of small producers by buying working interests in production at bargain prices. Because proven reserves are purchased, the risks are much less than those associated with drilling funds. Oil and gas tax-shelter investments are also weighed relative to other investments, and as a result some real es-

---

1. Hughes, *Boomtown Recession*, Austin American-Statesman, Mar. 13, 1983, at 15, sec. C. The present rig count is still more than twice that of a decade ago when the count was approximately 1,000.

2. Green, *Taking Stock of Oil and Gas Income Funds*, NAT'L TAX SHELTER DIG., Mar. 1983, at 25–28.

tate investments are being considered more favorably than drilling or income funds because of the accelerated write-offs provided by the new depreciation schedules.

The result of this decline in drilling fund activity should be short-lived as oil prices stabilize and begin to rise again as most forecasters are predicting they will and as the dramatic drop in drilling costs spurs renewed exploration.[3] Given increased drilling fund activity, two positive things should result from the 1982 slump. First, drilling funds may be structured better for investor return in order to compete with alternative investments. A 1981 Pennsylvania fund[4] offered 120 percent payout to subscribers before the sponsor participation in revenue increased from 1 percent to 25 percent, and this sort of arrangement may be prescient. Second, the sponsors who survive the 1982 downturn should be overall a financially sounder group than existed when prices were soaring and even marginal funds with inexperienced sponsors could make money. This factor will result in more investor confidence in oil and gas investments.

Should a person invest in an oil and gas drilling fund? Competition for investors is much more intense, and as a result investors have more well-structured funds from which to choose. A penchant for gambling—that oil or gas will be found and that prices will increase—is still required for an investor to select a drilling fund over more mundane tax shelters.

---

3. *Id.*
4. Omni 1981 Oil and Gas Programs.

Statutes and Rules Cited

[References are to sections in the text.]

UNITED STATES

United States Code

United States Code—*continued*

Crude Oil Windfall Profit Tax Act of 1980
(Public Law No. 96–223)

Economic Recovery Tax Act of 1981
(Public Law No. 97–34)

Natural Gas Policy Act

Securities Act of 1933
(Public Law No. 73–22)

Securities Exchange Act of 1934
(Public Law No. 73–291)

Small Business Investment Incentive Act of 1980
(Public Law No. 96–477)

Subchapter S Revision Act of 1982
(Public Law No. 97–354)

Tax Equity & Fiscal Responsibility Act of 1982
(Public Law No. 97–248)

Tax Reduction Act of 1975
(Public Law No. 94–12)

Tax Reform Act of 1969
(Public Law No. 91–172)

Code of Federal Regulations

Code of Federal Regulations—*continued*

Treasury Regulations

Revenue Rulings

Revenue Procedures

Securities & Exchange Commission Rulings

Securities & Exchange Commission Rulings—*continued*

MISCELLANEOUS

Uniform Limited Partnership Act

Uniform Limited Partnership Act (1976)

Uniform Partnership Act

TEXAS

Texas Revised Civil Statutes

Texas Administrative Code

Title 7

Texas Natural Resources Code

Texas Uniform Limited Partnership Act

Texas Uniform Partnership Act

Texas Attorney General Opinion

Cases Cited

[References are to sections in the text.]

Subject Index

[Decimal reference numbers are to sections and tables and figures (where indicated) in the text. This index does not cover citations to specific rules and sections of acts except where there is significant subject-matter treatment of specific regulations and acts. For citations to specific rules and sections of acts, see Statutes and Rules Cited.]

Q

R

A NOTE ON PRODUCTION

The typeface of this book—11 on 14 for the text and 9 on 11 for the footnotes—is ITC Garamond Book, composed on a Linotron 202, with the cover in Ludlow Garamond, by G&S Typesetters, Austin, Texas. The text paper is Weyerhaeuser 60-lb. Cougar opaque white, and the cover material is Sturdite Spanish grain imitation leather. Illustrations are by Gracie Barbé. Printing is by Hart Graphics, Austin, and binding by Custom Bookbinders, Austin.